NEC4: The Role of the *Project Manager*

Other titles in the Roles and Responsibilities series:

NEC4: The Role of the Supervisor (2017)
B. Mitchell and B. Trebes. ISBN: 978-0-7277-6355-6

NEC4: The Role of the *Project Manager*

Bronwyn Mitchell and Barry Trebes

Published by ICE Publishing, One Great George Street, Westminster, London SW1P 3AA

Full details of ICE Publishing representatives and distributors can be found at:
www.icebookshop.com/bookshop_contact.asp

Other titles by ICE Publishing:

NEC4: The Role of the Supervisor
B. Mitchell and B. Trebes. ISBN: 978-0-7277-6355-6
NEC4 Practical Solutions
R. Gerrard and S. Kings. ISBN: 978-0-7277-6199-6
Managing Reality, A practical guide to applying NEC4, Third Edition. 5-volume set.
B. Mitchell and B. Trebes. ISBN 978-0-7277-6195-8

www.icebookshop.com

A catalogue record for this book is available from the British Library

ISBN 978-0-7277-6353-2

© Thomas Telford Limited 2018

ICE Publishing is a division of Thomas Telford Ltd, a wholly-owned subsidiary of the Institution of
Civil Engineers (ICE).

Commissioning Editor: Michael Fenton
Production Editor: Madhubanti Bhattacharyya
Market Development Executive: Elizabeth Hobson

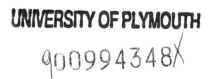

Typeset by Academic + Technical, Bristol
Index created by Pierke Bosschieter
Printed and bound in Great Britain by Bell and Bain, Glasgow

Contents

About the authors   ix
Foreword   xi
Abbreviations and other writing conventions used in this book   xiii

Section 1   **1**

01 .... **Introduction**   **3**
Overview of this book   3

Section 2   **5**

02 .... **Overview of the *Project Manager* role** **7**
2.1. Who is the *Project Manager*?   8
2.2. The role of the *Project Manager* in the ECC   8
2.3. *Project Manager* behaviours   12

Section 3   **15**

03 .... **Prior to contract award**   **17**
3.1. Contract strategy   17
3.2. Tender evaluation   17
3.3. Things to look for in the contract documentation   18
3.4. Things to look for in the Scope and Contract Data   18

Section 4   **19**

04 .... **After contract award but before starting on Site** **21**
4.1. Project team set-up   21
4.2. What the *Project Manager* is expected to do   23
4.3. Design that takes place prior to the first *access date*   23
4.4. The first Accepted Programme   24
4.5. Working Areas   24
4.6. The kick-off meeting with the *Contractor*   25

Section 5   **27**

05 .... **Starting on Site**   **29**
5.1. Overview   29
5.2. Checklist   29
5.3. The project team   30
5.4. Reporting and management   30
5.5. Ongoing design and design of Equipment   31
5.6. Subcontracting   32
5.7. Quality management   33
5.8. The Early Warning Register   36
5.9. The Accepted Programme   38

Section 6   **43**

06 .... **Chronological management procedures: first month/period** **45**
6.1. Regular meetings   45
6.2. First programme   45
6.3. First assessment of the amount due   46
6.4. First early warning notification   49
6.5. First compensation event   49
6.6. First marking of Equipment, Plant and Materials   55
6.7 First proposed instruction   55

Section 7   **57**

07 .... **Chronological management procedures: second and later months** **59**
7.1. Second and later (revised) programmes submitted by the *Contractor*   59

7.2. Assessment of the amount due: later months   60
7.3. Tests and inspections   63
7.4. Quality management   63
7.5. Dispute resolution   64

Section 8   **67**

08 ....
Completion and take over   **69**
8.1. Reaching Completion   69
8.2. Consequences of Completion   70
8.3. Assessment at Completion   71
8.4. Take over   71

Section 9   **75**

09 ....
After Completion and take over   **77**
9.1. Defects   77
9.2. Use of the *works*   78

Section 10   **79**

10 ....
After the *defects date*   **81**
10.1. The *defects date*   81
10.2. The Defects Certificate   81
10.3. Assessment of the final amount due   82

Section 11   **83**

11 ....
Post-project evaluation and learning from experience   **85**
11.1. Post-project evaluation   85
11.2. Learning from experience   86

12 ....
Summary   **87**

Appendices   **89**

Appendix 1 ....
Contract communications   **91**

Appendix 2 ....
Checklist for what the *Project Manager* needs to know from the *Client*   **99**

Appendix 3 ....
Checklist for what the *Project Manager* can look for in the contract documentation   **103**

Appendix 4 ....
Checklist for examples of things to look for in the Scope and Contract Data   **105**

Appendix 5 ....
Agenda for the first meeting between the *Project Manager* and the *Supervisor*   **109**

Appendix 6 ....
Checklist for the kick-off meeting with the *Contractor*   **111**

Appendix 7 ....
Checklist for templates that can be used by the *Project Manager*   **115**

Appendix 8 ....
Regular meeting agendas   **119**

Appendix 9 ....
Quality management considerations   **121**

Appendix 10 ....
Checklist for the *Project Manager* for managing quality   **123**

Appendix 11 ....
Agenda for a meeting about Defects and other matters about quality   **125**

Appendix 12 ....
Early warning meeting agendas   **127**

Appendix 13 ....
Checklist for sources of information to be included in the first programme   **129**

Appendix 14 ....
Checklist for information to be included in the first programme   **131**

Appendix 15 ....
Checklist for information to collect before undertaking the first assessment of the amount due   **135**

Appendix 16 . . . . Checklist for the first assessment of the amount due 137

Appendix 17 . . . . Checklist for early warnings 141

Appendix 18 . . . . Agenda for the monthly meeting between the *Project Manager* and the *Supervisor* 143

Appendix 19 . . . . Checklist for information to be included in subsequent programmes 145

Appendix 20 . . . . Checklist for information to be included in the *Contractor's* submitted programme 149

Appendix 21 . . . . Checklist for information to collect before undertaking assessments of the amount due 153

Appendix 22 . . . . Checklist for assessments that take place outwith core clause section 5 155

Appendix 23 . . . . Checklist for information to collect before undertaking the assessment of the amount due after Completion 157

Appendix 24 . . . . Checklist for assessment of the amount due after Completion of the whole of the *works* 159

Appendix 25 . . . . Checklists for Completion and take over 161

Appendix 26 . . . . Checklist for the *defects date* and the issue of the Defects Certificate 165

Appendix 27 . . . . Information that the *Project Manager* needs to consider for the final assessment 167

Appendix 28 . . . . Checklist for the final assessment of the amount due 169

Index 171

About the authors

Bronwyn Mitchell BCom(Hons), BProc, MBA, MCIPS, MCIArb
Bronwyn has been working with the NEC since 1995. She drafted an NEC website, which she sold to Thomas Telford Ltd in 1996, and assisted in redrafting a Small Works Contract, also written under licence, for Scottish contracting situations that was fundamentally adopted and later published as the Short Contract. She has worked with the ECC, ECSC, TSC, TSSC, PSC and PSSC as an advisor and trainer, providing guidance to both Parties and is the co-author of *Managing Reality*, a five-book series on the practical use of the ECC, and *NEC4: The Role of the Project Manager*.

Barry Trebes BSc(Hons), MSc, FRICS, FAPM, FInstCES
Barry has over 30 years' experience of providing consultancy, advice, facilitation and training on major developments across numerous business sectors, including aviation, defence, power, utilities, infrastructure, property, rail, highway and water, both in the UK and internationally. Barry provides NEC training and education, including NEC Masterclasses and Project Manager Accreditation courses provided by the Institution of Civil Engineers. He actively contributes to industry knowledge through numerous articles, and is the co-author of the *Managing Reality* series, *NEC4: The Role of the Project Manager* and BS PD 6079-4, 'Project management – Guide to project management in the construction industry'. He also initiated and helped to develop the first web-based NEC contract management system, named CCM, with MPS Ltd in 2000, and was a member of the NEC4 drafting team.

Foreword

I was delighted to be asked to write this foreword because of both the authors of this book and its subject. The *Project Manager* has a key role in the management of the NEC4 Engineering and Construction Contract and in contributing to meeting the *Client's* objectives on a project.

The main aim of this book is to guide *Project Managers* (or potential *Project Managers*) through their responsibilities. It covers what they should and should not do in undertaking the role effectively. This not only includes a good understanding of the specific terms of the contract but deals with other areas that the *Project Manager* might become involved in, and also sets out the necessary cultural aspects of being a competent *Project Manager*.

The *Project Manager* is one of the few named roles in the contract, and most people will naturally compare this to a more traditional Engineer/Contract Administrator/role. These roles are different, and it's important that this is understood. The role of *Project Manager* is extremely wide ranging and proactive – technical of necessity, but practical too. People acting in this role should get a good understanding of its implications before they undertake it.

There are no other substantial texts that deal with the role of the *Project Manager*, so the benefits and importance of this book are clear. It is helpful that the book is authoritative and well laid out in a chronological order. This book is also very relevant to supporting the NEC series of accreditations for the various roles under the NEC contracts. More details on the ECC Project Manager Accreditation can be found on www.neccontract.com, and I would urge anyone acting in that role, or clients wishing to appoint someone in that role, to give consideration to getting accredited or using accredited *Project Managers* on their projects.

Following on from their highly credible 'Managing Reality' series of books, Bronwyn Mitchell and Barry Trebes have co-written this book in their own straightforward and incredibly helpful style. The extensive experience of NEC contracts that they both have is obvious, and I cannot think of two better authors to have written this book.

Better and more informed *Project Managers* can only contribute to a better industry. I hope that you will enjoy this book as much as I did.

Robert Alan Gerrard
Secretary, NEC Users' Group

Abbreviations and other writing conventions used in this book

Abbreviations

CD1	Contract Data part one
CD2	Contract Data part two
ECC	NEC4 Engineering and Construction Contract
HGCR Act	Housing Grants, Construction and Regeneration Act 1996
ICE	Institution of Civil Engineers
M&E	mechanical and electrical
NEC4	New Engineering Contract (fourth edition)
SCC	Schedule of Cost Components
SSCC	Short Schedule of Cost Components

Other writing conventions

- This book is designed as an effective companion to the NEC4 *Project Manager* training courses available through ICE, including ECC *Project Manager* Accreditation. It provides complementary information and, as such, much material is taken directly from the ECC.
- ECC Guidance Notes are published in volumes 1 to 4 by ICE.
- Initial capitals indicate defined terms in the ECC, while italics indicate identified terms in the Contract Data.

Section 1

NEC4: The Role of the *Project Manager*
ISBN 978-0-7277-6353-2

Introduction

The fourth edition of the New Engineering Contract (NEC4) – like its earlier versions – is specifically designed to promote and stimulate good practice in the management of projects. In the NEC4 Engineering and Construction Contract (ECC) the *Project Manager* has a key role in ensuring that the good management practice that underpins the contract is realised in practice. The other key people (named in the contract) are the *Supervisor*, the *Client*, the *Contractor* and the *Adjudicator*.

As the role title would indicate, the *Project Manager* is the person who project manages the ECC *works* project on behalf of the *Client*. The *Project Manager* may be an employee of the *Client* or they may be contracted with the *Client* using, for example, the NEC4 Professional Services Contract. The *Project Manager* manages the contract for the *Client* with the intention of meeting the *Client's* objectives. The *Project Manager* can add value to the project right from the outset in a number of ways, for example by providing advice on the procurement of design, on estimates of costs and time, on options or alternative solutions, or on choosing the most appropriate contracting strategy.

The role of the *Project Manager* is clearly described in the ECC, but there are also areas of communication and interaction that are assumed or on which the ECC is silent. One of the objectives of this book is to describe in more detail the *Project Manager's* actions in these areas.

Overview of this book

This book provides guidelines to those who have been allocated the role of *Project Manager* on an ECC contract by describing the obligations, activities and culture necessary to fulfil the role effectively. The book is designed to be an effective companion to the Institution of Civil Engineers (ICE) NEC4 accreditation course on the role of the *Project Manager*, and provides complementary information.

The book describes the technical and behavioural traits that will be of use to the *Project Manager*, and is divided into a chronological set of learning outcomes, as shown in Figure 1.1, that mirror the order in which a project can be established and implemented.

Figure 1.1 *Project Manager* book structure

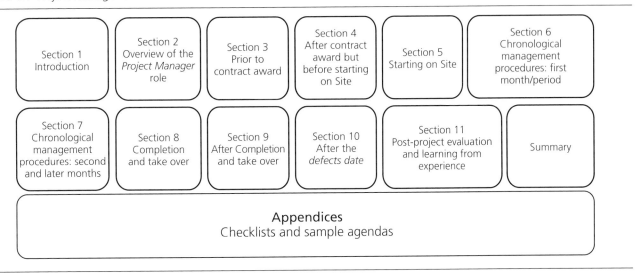

Section 2

NEC4: The Role of the *Project Manager*
ISBN 978-0-7277-6353-2

Overview of the *Project Manager* role

The *Project Manager* should be proactive in the management of their actions so that they bring the contract to life, and should put in place working practices that enable them to successfully undertake their role in the contract (Table 2.1). They interact with other people every day, including the *Supervisor*, the *Contractor*, Subcontractors, the *Client*, and day visitors to the Site (e.g. inspectors), therefore good interpersonal skills are required to be a success in the role of *Project Manager*.

It is worth emphasising that successful projects don't just happen: they are reliant on people carrying out their contractual roles and the processes required by the contract. The ECC describes the process, but people make the difference, and the *Project Manager* plays a key role in providing leadership, encouraging a collaborative culture and working with the wider project team.

Table 2.1 The main duties of the *Project Manager*

General duties	Programme	Change management	Payment	Risk management
Act as stated in the contract Act in a spirit of mutual trust and co-operation	Monitor the *Contractor's* planned Completion and Completion Date	Manage the compensation event process so that the *Contractor* is fairly compensated for any *Client*-initiated change on the project	Assess payment after each *assessment interval*	Use and encourage the *Contractor* to use the early warning procedure to identify and manage risk
Communicate and issue documents as required by the ECC	Review and accept programme submissions	Ensure the *Contractor* notifies compensation events in good time so that the *Client* is not disadvantaged		

Based on the project management triangle (Figure 2.1), the ECC describes procedures that the *Project Manager* and Others can use to manage the contract effectively, including

- risk management through the use of the Early Warning Register and the early warning procedure
- change management through the compensation event procedure.

Figure 2.1 The project management triangle

2.1. Who is the *Project Manager*?

The *Project Manager* is one of the named roles in the ECC. The other named roles (see Figure 2.2) are

- the *Client*
- the *Contractor*
- the *Supervisor*
- the *Adjudicator*.

Where the *Project Manager* is named:

- The *Project Manager* is identified in the Contract Data. In some cases, the Contract Data may simply state the name of an organisation to which the *Client* is outsourcing the role of the *Project Manager*; however, it is important that the *Client* approves the person assigned the role and that the *Client* gives the name and contact details of the *Project Manager* to the *Contractor* prior to the *starting date* of the contract.

Appointing a *Project Manager*:

- The sooner the *Project Manager* is appointed and starts working with the rest of the team and with the documents that are relevant to the project, the better their knowledge and the more effective their contribution will be. A *Project Manager* who has been involved in a project since its business need was first identified will be in a better position to add value than one who is introduced post-commencement.

Replacing the *Project Manager*:

- The *Client* may replace the *Project Manager* after it has notified the *Contractor* (clause 14.4).

> The *Project Manager* is one of the named roles in the ECC, and is required to carry out specific duties under the contract in relation to time, risks, change and payment.

Figure 2.2 The role of the *Project Manager*

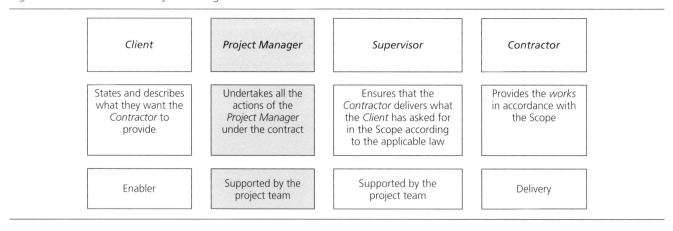

2.2. The role of the *Project Manager* in the ECC

This section provides an overview of the role of the *Project Manager* and also the framework for the detailed discussions in other sections in the book.

2.2.1 Clause 10.1

Clause 10.1 states that the *Project Manager* '**shall act as stated in the contract**' and clause 10.2 provides that the *Project Manager* acts '**in a spirit of mutual trust and co-operation**'.

There have been endless discussions about 'mutual trust and co-operation', as well as numerous law articles and expert opinions. In the end, the project team must decide together what this means for their project, and the most obvious way to do this is through the actions taken by the individual members of the project team.

Throughout this book, areas where the *Project Manager* is acting as stated in the contract and/or in a spirit of mutual trust and co-operation will be indicated: for example, see Section 2.3 below.

2.2.2 Role of the *Project Manager* differentiated from the *Supervisor*

The *Project Manager* does not abdicate their responsibility for the project's quality outcome just because the *Supervisor* is involved in carrying out or watching the tests and inspections described in the Scope and in notifying Defects. The *Project Manager* is still responsible for ensuring that the *works* achieve the *Client's* objectives, and that means that it is the *Project Manager* who is responsible for deciding the date on which Completion is achieved, the final out-turn cost to the *Client* and the quality of the *works*. The *Supervisor* must confine their actions to the scope of the services described in their personal contract and the requirements of the Scope, but the *Project Manager* can accept Defects and instruct changes to the Scope, and they are therefore just as actively involved in the quality of the *works* as the *Supervisor* (Figure 2.3).

> The *Project Manager* works with the *Supervisor* to check that the quality of the *works* required by the Scope is adhered to by the *Contractor*.

The *Supervisor* is not required to report to the *Project Manager* in a direct-line-reporting manner, and the *Supervisor* carries out their duties independently of the *Project Manager*. However, the *Project Manager* will want to work closely with the *Supervisor*, to understand whether the *Contractor* is providing the quality of the *works* required and whether the time and/or cost of the project is being impacted negatively by Defects and/or failed tests or inspections.

Figure 2.3 Role differentiation in the *Client's* organisation

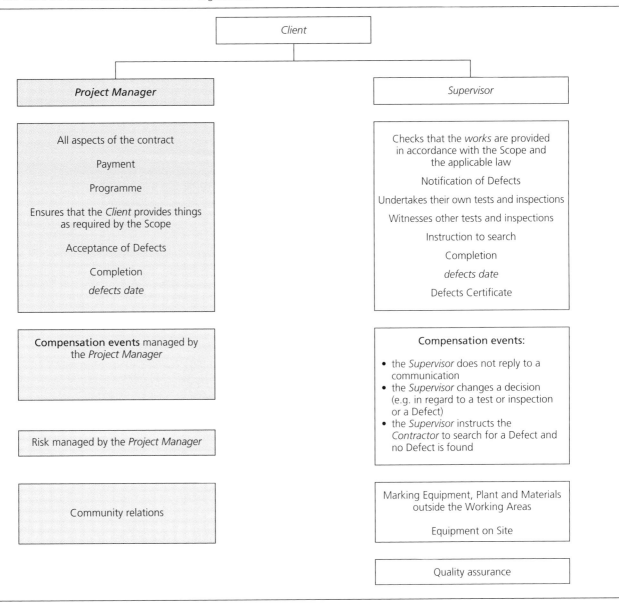

2.2.3 Working through other people

Of course, there will be communications, discussions and notifications between the *Project Manager* and the other people involved in the contract, such as the *Client* and the *Supervisor*. Indeed, the contract would not work well if they did not talk to each other or worked in isolation from each other. However, the relationships between these persons are governed more by their professional contracts (internal or through a professional contract such as the NEC4 Professional Services Contract) than by the ECC.

The *Project Manager* may delegate parts of their role to other professionals, or the *Client* may require delegation by the *Project Manager* to other team members, according to their area of expertise. Such roles could include a cost advisor (core clause section 5), a programmer/planner (core clause section 3), a change manager (core clause section 6) and a designer (core clause section 2).

2.2.4 Interactions of the *Project Manager*

Contractually, the *Project Manager* is required to interact with the *Contractor*, the *Client* and the *Supervisor* (Figure 2.4). However, there are many other people involved in the contract – that is, people other than the named parties in the Contract Data – with whom the *Project Manager* may be required to interact or whom the *Project Manager* may come across at the Site. Examples are

- anyone to whom the *Project Manager* has delegated actions, for example
  - a quantity surveyor (payment)
  - a planner (the programme)
  - a risk analyser (the Early Warning Register)
  - an administrator (early warnings and compensation events)
- the *Supervisor* (or more than one where specialisms have been shared among several people)
- health and safety inspectors or specialists
- environmental inspectors or specialists
- building control officers
- community relations people
- insurance inspectors
- anyone on the design team for a design-and-build-style contract
- other officials, such as planning officers
- utility providers
- Subcontractors and suppliers of Plant and Materials delivered to the Working Areas.

Figure 2.4 Interactions of those involved in the project

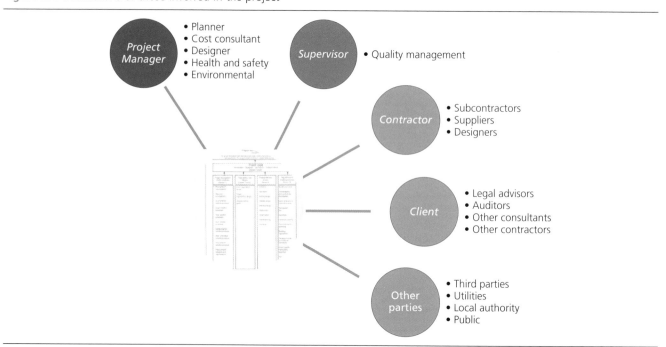

2.2.5 Communications

The ECC requires the *Project Manager's* communications to follow certain rules, and these rules require professional record-keeping by the *Project Manager*.

The *Client* can choose to describe a communication system in the Scope. If this is the case, then clause 13.2 provides that a communication has effect only when it is communicated through the communication system specified in the Scope. This means that all those working on the contract must be aware of and be able to use the communication system if they are to ensure that their communications have effect.

The use of a communication system helps to set expectations about communications and recognises that, in general, communications are more effective if a standard is created and used across the whole project. The *Client* may have a communication (software) system that the project members are required to use, or even a simple database available to most computer users.

Whatever method is chosen, it is recommended that a series of pro-forma communications is set up to be used on the project, where each pro forma

- has a unique identifier (e.g. Project xxx/EW/001)
- states the relevant clauses to which the communication relates
- refers to timescales to facilitate timely communication by all parties
- contains a set distribution list for each type of communication to facilitate contractually correct communication to all those expecting to receive notifications.

It may be that the *Project Manager* will provide direction on the format and style of project communications. They should be able to assess the system and whether it will work, checking that

- the format of communications can be used by the whole project team
- the recording system allows easy access to information
- project records are uniform and easily auditable.

A communication system described in the Scope could cover areas such as

- the language to be used (as a reiteration of the *language of the contract* identified in Contract Data part one (CD1))
- how contractual communications will take place (e.g. by email or using a software system that automatically sends the communication to the addressee) so that all parties know when their communication will take effect
- basic information to include in each communication, such as from whom, to whom, the date, the clause reference, and the type of communication (e.g. notification, certificate or reply)
- a set distribution list for all communications, to ensure that all parties are aware of contractual events
- specific days for communications (e.g. all notifications are issued at the weekly site meeting on Monday morning).

In any case, the *Client* should ensure that the relevant communications protocol is included in the scopes of services for the *Project Manager* and the *Supervisor* (whether outsourced or not) and, more importantly for the purposes of this book, in the Scope of the project so that all parties are using the same protocol.

The following lists the communication protocol required by the ECC:

- all communications (e.g. instructions, certificates, replies and notifications) from the *Project Manager* must be in a form that can be read, copied and recorded (clause 13.1)
- all communications must be in the *language of the contract* (e.g. English) (clause 13.1)
- The *Project Manager* must issue their certificates to the *Client* and the *Contractor* (clause 13.6)
- *Project Manager* notifications and certificates required by the contract must be communicated separately from other communications (clause 13.7)
- where the *Project Manager* is required to reply to a communication, they must do so within the *period for reply* (clause 13.3).

It is worth emphasising that notifications required by the contract **must** be communicated separately from other communications. The rules for communications and the list of communications required from the *Project Manager* in an ECC contract are listed in Appendices 1A and 1B in this book, respectively.

2.2.6   Communications with the *Client*

The *Client* is required to undertake activities and provide things at certain points in the contract. The *Project Manager* may be required to manage this interface on behalf of the *Client* and keep them informed of any changes to the dates and timing of when the *Client's* actions may be required. As with the *Supervisor*, the *Client* may want to describe reports and interactions in the *Project Manager's* personal contract so that the *Project Manager* keeps the *Client* informed with updates on the project.

In addition, the *Project Manager* needs to be clear about what they can do without permission or conference with the *Client*. An instruction changing the Scope could impact on the Prices – does the *Project Manager* have authority to commit the *Client* to increases in cost without first describing the problem and the solution to the *Client*? Prior to starting work on the project the *Project Manager* may wish to go through the actions that the ECC requires of them with the *Client* so that both are fully aware of the parameters that delimit their role. These parameters should then be clearly set out in the *Project Manager's* contract.

A checklist showing what the *Project Manager* needs to know from the *Client* about their role is included in Appendix 2.

2.3.   *Project Manager* behaviours

The ECC provides opportunities for the *Project Manager* and the *Contractor* to work together for the mutual benefit of the *Client* and the *Contractor*. An effective *Project Manager* will possess and be able to use soft skills and behavioural interactions that enhance the project outcomes. Apart from the obvious 'mutual trust and co-operation', the ECC provides for

- effective, timely communications (using the *period for reply*, and clauses that require responses to communications, e.g. clause 31.1)
- joint decision-making (e.g. as used in clause 15.3)
- objective decision-making (e.g. clauses 60.1(12) and 60.1(13))
- solutions if the *Project Manager* does not do something that they are required to do (e.g. clause 64.4)
- sanctions for not doing things that are required (e.g. clause 61.5)
- ways to keep things moving even in uncertainty (e.g. clauses 61.6 and 62.5).

How should the *Project Manager* act in the following situations?

- While assessing the *Contractor's* quotation for a compensation event, the *Project Manager* notices that the *Contractor* has made an error in calculating the forecast Defined Cost and that the quotation is £4500 less than the Defined Cost is likely to be. Should the *Project Manager*
  - notify their acceptance of the *Contractor's* quotation (clause 66.1)
  - assess the compensation event themselves (clause 64.1) and make the correction or
  - instruct the *Contractor* to submit a revised quotation (clause 62.3) after informing the *Contractor* of the error?
- The *Project Manager* realises that if the *Contractor* changes a planned method of working (in an ECC main Option A, with secondary option X6 (bonus for early Completion)), the *Contractor* will be able to complete the *works* before the Completion Date and will therefore benefit with a bonus for early Completion. Should the *Project Manager*
  - do nothing
  - explain the change to the *Contractor* and suggest that they use clause 55.3 and submit a revised Activity Schedule and programme for acceptance or
  - notify an early warning (clause 15.1)?
- The *Project Manager* realises that the *Contractor's* inexperience with medical gases has led to an underestimate of time in the Accepted Programme. The *Project Manager* had not previously noted that the resource statement in the Accepted Programme is unrealistic. Should the *Project Manager*
  - do nothing
  - notify an early warning (clause 15.1) or
  - instruct the *Contractor* to submit a revised programme (clause 32.2)?

Collaborative working (Table 2.2) is much more efficient and effective for everyone involved in the contract. If the *Project Manager* thinks of the *Contractor* as an expert, and recognises that the *Project Manager* is not the only one who can find solutions, then they will not be uncomfortable about consulting the *Contractor* about the best way to move the project forward.

Table 2.2 Good and bad habits of a *Project Manager*

Good habits	Bad habits
■ Acting as stated ■ Working in a spirit of mutual trust and co-operation ■ Actively communicating, within the *period for reply* or as otherwise stated in the contract ■ Making decisions and giving reasons for those decisions ■ Working collaboratively	■ Not acting as stated ■ Not giving early warnings ■ Not giving instructions ■ Not notifying compensation events ■ Not communicating ■ Not discussing or consulting ■ Inspecting Defined Cost late ■ Not working with the *Supervisor*

If your first instinct is to baulk, posture, give orders and react, try instead to listen, respond in a positive way, be open to new ideas, collaborate and participate. You might not get it right every time but the project will move forward, which will benefit all those involved.

In conclusion, the *Project Manager* should be proactive in the management of their actions so that they bring the contract to life. They should put in place working practices that will enable them to undertake their role in the contract successfully. Good interpersonal skills are required for success in the role of *Project Manager*, as it will require daily interactions with others, such as the *Supervisor*, the *Contractor*, Subcontractors, the *Client* and day visitors to the Site (e.g. inspectors). Successful projects don't just happen; they are reliant on people carrying out their contractual roles and the processes required by the contract. The ECC describes the process, but people make the difference, and the *Project Manager* plays the key role in behaving in a collaborative culture and working with the wider project team.

Section 3

NEC4: The Role of the *Project Manager*
ISBN 978-0-7277-6353-2

ICE Publishing: All rights reserved
http://dx.doi.org/10.1680/nectrpm.63532.017

Prior to contract award

If the *Project Manager* is involved in the project prior to contract award, they will find it helpful to become familiar with the contract documents, including the *Contractor's* tender, so that they understand what the *Client* expects and the constraints within which the *Contractor* is required to work.

3.1. Contract strategy

The contract strategy is decided by the *Client* prior to the issue of procurement documents. It will determine aspects of the project such as

- how the *Contractor* will be paid (mains Options A–F)
- whether part of the amount due will be held by the *Client* as retention (Option X16)
- whether the *Contractor's* liability is limited (Option X18).

Unless the *Client* involves the *Project Manager* in the project right from the beginning, the *Project Manager* will have no input into the contract strategy and no opportunity to include their experience and knowledge in the contract.

The *Project Manager* should go into the project knowing what the contract strategy is and the impact of the chosen Options on the programme, the cost to the *Client* and the required quality and performance.

3.1.1 Options

The Options chosen as part of the contract strategy and listed in CD1 by the *Client* should be reflective of the Scope and, to a lesser extent, the Site Information. The *Project Manager* should be familiar with all the Options chosen and how they could affect the project.

3.1.2 Meaning of optional statements

The optional statements included in CD1 by the *Client* also affect the way that the *Project Manager* will approach the project. For example, if the *Client* has indicated that it is not willing to take over the *works* before the Completion Date, the *Project Manager* will understand that

- if the *works* are ready before the Completion Date, the *Client* is not obliged to take over at that point but can wait until after Completion is certified by the *Project Manager*
- Option X6 (bonus for early Completion) is not appropriate as part of the contract strategy
- there is probably no need for a reason in the Scope for the *Client* to use part of the *works* before Completion has been certified.

3.2. Tender evaluation

If the *Project Manager* is fortunate enough to be a part of the tender evaluation team, they will be able to review the tender submissions. This will give them the opportunity to get to know more about the tenderer who will become the *Contractor*, before work starts on the Site.

> If you were the *Project Manager*, which document would you choose to look at first: the programme tendered or the Prices?

The programme and (to a more limited degree) the Activity Schedule will provide information about how the *Contractor* plans to approach the *works*. The programme will also give insight into the *Contractor's* approach to risk.

The degree of synchronicity between the two documents may provide information about how realistic the Activity Schedule actually is.

3.2.1 Programme

The *Client* is not required to ask for a programme to be submitted with the tender. In some circumstances, it might choose to request an outline programme rather than a complete ECC programme: the drafting of the first programme can be expensive to the tenderers in terms of resources and time; and the *Client* must have provided enough information to the tenderers for them to be able to provide a statement of resources and how the tenderer plans to provide the *works*. Another option is for the *Client* to ask for a fully compliant ECC programme for chosen activities, which can then be evaluated across tenderers and scored.

The *Project Manager* can use the tender programme to mark out some of their activities so that they and the *Contractor* can use this at the kick-off meeting to agree, among other things, dates for weekly/monthly meetings and for assessment of the amount due. The *Project Manager* can also choose to ask for parts of the programme in an ordered way, rather than as one programme in a couple of weeks' time, so that the programme can be reviewed as it is produced, especially if it is a long and complicated project.

3.3. Things to look for in the contract documentation

When first reading the documents that comprise the contract, the *Project Manager* will look for information, dates and details that will help them produce an agenda for the kick-off meeting and a list of items to address at monthly meetings, especially where they have spotted anomalies or potential weaknesses. Each project will be different in terms of complexity, timelines and what aspects are important to the *Client* at the time.

An example checklist of what to look for in the contract documentation is included in Appendix 3.

3.4. Things to look for in the Scope and Contract Data

Many ECC clauses contain references to the Scope or the Contract Data and so, for those clauses to work properly, the expected information in the Scope or the Contract Data needs to be completed. The reader must be able to find the information referred to as being in the Scope and Contract Data. In particular, the *Project Manager* needs to understand what their *period for reply* is (clause 13.3) and what the Scope requires the programme to include (clause 31.2).

A checklist of things that the *Project Manager* can find in the Scope and Contract Data and that affect their role is included in Appendix 4.

Section 4

NEC4: The Role of the *Project Manager*
ISBN 978-0-7277-6353-2

After contract award but before starting on Site

The *Client* may have set up their whole project team prior to contract award to allow the agreement of project procedures among the team. However, it is more likely that the creation of the team will take place after the contract has been awarded to the *Contractor*. The *Project Manager* will want to meet their team and arrange a pre-Site meeting so that the whole team is aware of the various roles and who are the go-to people in the contract.

4.1. Project team set-up

The project team set-up is decided by the *Client*. There are several different ways in which a successful team can be organised, so the *Project Manager* needs to understand their project reporting lines so that they can communicate efficiently with the *Client*, Others and their own team in order to keep the project on track.

The *Project Manager* must bear in mind that the role of the *Client* and Others can be carried out by more than one person. For example, the person to whom the *Project Manager* speaks about work to be done by the *Client* according to the Accepted Programme (clause 31.2) could be different from the person who provides access to the *Contractor* after take over (clause 44.4) and different from the person who effects payment to the *Contractor* (clause 51.2).

The *Project Manager* needs to be aware of all the people with whom they are required to interact in the project. In a small project, the *Project Manager* may carry out all the actions required by the *Project Manager* and also those of the *Supervisor* role as well. In a larger project, the *Project Manager* may have an overseeing role, having delegated the actions required by core clause section 5 to a quantity surveyor, the actions required by core clause section 3 to a planner and the actions required by core clause section 4 to the *Supervisor*.

4.1.1 Delegation

The *Project Manager* can use clause 14.2 to delegate any of their actions to other people. A template form for notification is included in the ECC Guidance Note.

The delegation may be for a specific period of time, such as when the *Project Manager* is on holiday, for the duration of the project or for any other length of time, as long as the *Project Manager* has notified the *Contractor* first that the delegation is going to take place.

The *Client* may require the *Project Manager* to delegate some of their actions if the *Client's* organisation is set up in a certain way, for example if it separates the payment transactions from the programme transactions.

4.1.2 Agreed procedures

The *Client* may require certain project procedures to be in place to support the procedures required by the ECC.

For example, clause 62.3 requires the *Project Manager* to reply within 2 weeks to a quotation for a compensation event submitted by the *Contractor*. It may be that the *Client* has in place a company policy that requires any changes in the Prices to be approved by a project board and that the *Project Manager* does not have the *Client's* authority to accept changes to the Prices. In this case, the *Client* may provide for a longer period of response, using an Option Z clause to increase the 2-week response period, or may give the *Project Manager* the power to call a special approval board meeting for compensation events.

Similarly, there are points in the ECC where the *Client* is part of a procedure but there are no directions for the communications about the procedure. In such instances the *Client* may choose to make sure that the *Project Manager's* personal contract allows for these communications. For example, a compensation event in clause 60.1(14) is the occurrence of an event that is a *Client's* liability stated in the contract. Further to the compensation event procedure described in core clause section 6, it is probable that the *Client* would like to be made aware of the liability event. The *Client* may want to ensure, through the use of the *Project Manager's* personal contract, that the *Client* is kept aware of such events so that they can protect their interests. Section 2.2.6 in this book on communications with the *Client* refers to Appendix 2, which provides a larger list of items that the *Project Manager* may want to know.

4.1.3 Management of other project members
If the *Project Manager* has delegated some of their actions to other people, whether voluntarily or through the *Client's* chosen project team set-up, they will need to agree management procedures with them so that they are always aware of any changes to the project.

4.1.3.1 Quantity surveyor/cost consultant
A quantity surveyor could be delegated the role of the *Project Manager* to undertake the duties under core clause section 5 and other assessment clauses (e.g. clauses 25.2, 25.3, 41.6, 46.1, 46.2 and 93.1). Nevertheless, the communications between the *Project Manager* and their quantity surveyor will need to be frequent and planned so that both are aware of changes to the project (not only compensation events) and the impact on the Prices.

The *Project Manager* can either require certain procedures from the quantity surveyor or can ask the quantity surveyor (who may be a consultant and not an employee of the *Client*) how they intend to undertake their delegated duties. The *Project Manager* will want to know

- how the quantity surveyor will approach every assessment under clause 50.1 and all other assessment clauses
- what records the quantity surveyor wants available (e.g. clause 52.2)
- how and when the quantity surveyor wants to receive forecasts of the total Defined Cost (main Options C to F clause 20.4)
- how the quantity surveyor will approach the *Contractor's* application for payment under clause 50.2
- whether the quantity surveyor wants to schedule a regular time (e.g. weekly) with the *Contractor* to walk around the Site to gauge progress and/or go through events on Site and deal with assessments and forecasts in a topical fashion
- how the quantity surveyor will deal with compensation events.

The *Project Manager* may require the quantity surveyor's presence at

- early warning meetings
- weekly progress meetings, with or without other people (e.g. the planner and *Contractor*)
- meetings to discuss the programme, the Prices and Defects.

4.1.3.2 Planner/programmer
The management of the programme is at the heart of NEC contracts and so requires the appropriate level of resources. Some projects will require heavy input into the programme, which may result in the *Project Manager* delegating their programme duties to a planner/programmer on a full- or part-time basis. The ECC requirements for the Accepted Programme are comprehensive, and the assessment of submitted programmes and the charting of progress could be demanding and time consuming. On any day, the *Project Manager* will want to be able to look at the Accepted Programme and know exactly where the project is and why the programme has changed from its first acceptance.

The *Project Manager* will want to meet regularly with the planner, and may require their presence at

- early warning meetings
- weekly progress meetings, with or without other people (e.g. the quantity surveyor and *Contractor*)
- meetings to discuss the programme, the Prices and Defects.

4.1.3.3 *Supervisor*
There are few interactions between the *Supervisor* and the *Project Manager* that are dictated by the ECC. The *Project Manager* will want to meet regularly with the *Supervisor* and set up working relationship controls so that they are kept aware of Defects and other quality matters that may affect the programme and Prices and how the *Client* uses the *works*, and which may impact on Others and other people. In particular, the *Project Manager* will want the following from the *Supervisor*:

- weekly meetings to update the *Project Manager* on the list of Defects, test/inspections undertaken and any issues, such as a Defect shown by a test

- an agreed procedure on notification where, for example, the *Client* is required to provide materials, facilities and samples for an imminent test or for a repeated test
- an agreed procedure on notification where, for example, the *Project Manager* will need to assess the cost incurred by the *Client* in repeating a test
- attendance at early warning meetings.

The *Supervisor*, if in harness with the *Project Manager*, can be their eyes and ears in the Working Areas and can provide a very valuable source of information for the *Project Manager* that can help them make decisions in relation to compensation events, the revised programme, Defects, early warnings, Key Dates and so on.

The recording of this information will not happen by accident, and the *Project Manager* is well advised to discuss with the *Supervisor* the records and information that will be recorded. Appendix 5 identifies the types of items to be considered and how they interface with the role of the *Project Manager*.

> The *Supervisor*, if briefed, can act as the *Project Manager's* eyes and ears in the Working Areas.

4.1.3.4 Adjudicator
Although the *Client* is required to name the *Adjudicator* in CD1, more often than not the *Client* will prefer to name a body who will appoint an *Adjudicator* should a dispute get that far. The *Project Manager's* actions with regards to adjudication are dictated by the ECC, and, in many cases, it will be easier if the *Project Manager* is involved in the dispute only as much as they are required to be. The *Project Manager* will still have to deal with the *Contractor* on a day-to-day basis, which can be difficult with an adjudication in the background.

4.2. What the *Project Manager* is expected to do
The ECC is clear about the role of the *Project Manager*, but the *Project Manager* still has to interact with the *Client* and the *Supervisor*. Communication between the *Project Manager* and the *Client* and the *Supervisor* is not dictated by the ECC, and so the *Project Manager* will need to work out their own way of working with them so that all of them can fulfil their role under the ECC.

The *Project Manager* will want to get familiar with their personal contract with the *Client* so that they are aware of the parameters outlined by the *Client* (see Section 2.2.6 of this book). They will also want to meet with the *Client's* representative (sometimes a project sponsor) and the *Supervisor* and others in the team before the project kick-off meeting so that they can provide appropriate information to the *Contractor* about how the project is going to be run and managed.

4.3. Design that takes place prior to the first *access date*
The *Project Manager* will want to set up procedures and protocols for any design that the *Contractor* is required to undertake. The Scope should state the extent of the *works* that are to be designed by the *Contractor*. CD1 identifies the *Client's* Scope, and Contract Data part two (CD2) identifies the information for the *Contractor's* design.

In accordance with the ECC, the *Contractor* will design the parts of the *works* that the Scope states it is to design, and will submit the particulars of this design to the *Project Manager* for acceptance as required by the Scope. It may submit its design for acceptance in parts if the design of each part can be assessed fully. The *Contractor* cannot proceed with the work until the *Project Manager* has accepted the design. (The *Contractor* may be required to submit particulars of design on an item of Equipment if the *Project Manager* instructs them to do so.)

Therefore, the *Project Manager* should ask for the following information from the *Contractor*:

- Is the *Contractor* intending to submit the design to the *Project Manager* for acceptance in parts? Clause 21.3 does not require the *Contractor* to seek permission from the *Project Manager* before it submits the design in parts, and the two may have differing opinions on whether the design of each part can be assessed fully, so it is best if the *Project Manager* and the *Contractor* agree about the parts and the timing for submission before it happens.
- What is the programme for submitting the design? Has the *Contractor* left sufficient time in the programme for the *Project Manager's* review? This is particularly important if the contract is for a design-and-build solution.
- The *Contractor* should identify the timescales in which the *Project Manager* is required to reply to the design submissions (within the *period for reply*, clause 13.3), and the *Project Manager* should be reminded of this period, whether it is a specific period separately identified in CD1 by the *Client* or just the standard *period for reply*, or shown on the programme.

Since the *Contractor* cannot proceed with the work until the *Project Manager* has accepted the design (clause 21.2), the *Project Manager's* acceptance is critical to the programme of the *works*, as well as critical to the budget and performance. The communications and dialogue at this early stage of the contract affect the rest of the *works*, and so the *Project Manager* needs to be on top of the requirements of the Scope and what the *Contractor* needs in order to proceed.

The *Project Manager* should ascertain if there are other procedural requirements in the Scope or other reasons for not accepting the design.

4.4. The first Accepted Programme

If the *Contractor* was required to submit a programme with their tender (i.e. CD1 was silent on a first programme, and CD2 required the tenderer to identify a programme), then the *Project Manager* will want to become familiar with the programme, which, post-award, becomes the Accepted Programme.

The ECC provides a comprehensive list of what should be included in the programme (clause 31.2). The *Project Manager* should be familiar with this list and how it has been shown by the *Contractor* on its programme submitted for acceptance. Further information about the Accepted Programme and the *Project Manager's* duties with regards to the Accepted Programme are included in Section 5.9 of this book.

At the kick-off meeting with the *Contractor* it is worth the *Project Manager* checking to see how the *Contractor* is progressing with their first programme for acceptance if one has not been submitted with their tender in CD2. Rather than wait, the *Project Manager* could act in a spirit of mutual trust and co-operation and enquire mid-way through the submission period on how the *Contractor* is progressing and if there are any points or issues that it may wish to discuss.

4.5. Working Areas

The *Client* provides the *boundaries of the site* in their invitation to tender. The *Contractor* describes their *working areas* in their tender. As part of the tender evaluation, the *Client* will assess the areas proposed by the *Contractor* as *working areas* in CD2 and decide whether they (1) are necessary to Provide the Works and (2) will be used as part of the work in the contract. If the *Client* does not agree that parts of the *working areas* proposed fit in with the definition of the Working Areas, they should advise the *Contractor* of this decision prior to contract award so that the *Contractor* is aware that they may need to find other areas to use or change the way they plan to work if they cannot use the areas they had planned to use. Both Parties need to agree on what comprises the Working Areas, as these affect both the amount due to the *Contractor* and the movement of Equipment, Plant and Materials.

The Working Areas are

- the Site
- other areas used to Provide the Works, such as
  - the area to be used to make up reinforcement and formwork for the Site
  - the area to be used for the *Contractor's* site accommodation (e.g. offices, drying rooms and the canteen)
  - borrow pits
  - work beyond the boundary of the Site (e.g. the public highway).

Working Areas do not include

- the *Contractor's* head office, as the costs for this are recovered as part of the *Contractor's fee percentage*
- locations outside the Working Areas where Plant and Materials are manufactured or fabricated or where design work takes place, as stated in the Schedule of Cost Components (SCC) and Short Schedule of Cost Components (SSCC), cost components 6 (manufacture and fabrication) and 7 (design).

The Working Areas concept is important for the visibility and management of the *Contractor's* Defined Cost, as the *Contractor* is only paid for resources working within the Working Areas and certain cost areas outside the Working Areas as identified in the SCC and SSCC. Any costs incurred by the *Contractor* in areas outside the Working Areas do not form part of Defined Cost (unless included in cost component 6 or 7) and are not included in the amount due.

At the time of tender the *Contractor* will not always know what areas other than the Site they may require. The *Contractor* can use clause 16.3 to submit a proposal to the *Project Manager* to add an area to the Working Areas. The management of the Working Areas could be a topic to address in the monthly progress meetings.

4.6. The kick-off meeting with the *Contractor*

The kick-off meeting is not a requirement of the ECC, but it is essential to good project management. It may be the first time that the *Project Manager* meets the *Contractor* and the key people in the *Contractor's* interfacing team. It gives the *Project Manager* the opportunity to set out the communications methods and the contract procedures and requirements to the *Contractor* so that there is a mutual understanding from day 1.

Appendix 6 sets out the things that the *Project Manager* needs to consider at a kick-off meeting with the *Contractor*. Appendix 7 lists the communication forms that could be used by the *Project Manager* and the project team.

Section 5

NEC4: The Role of the *Project Manager*
ISBN 978-0-7277-6353-2

ICE Publishing: All rights reserved
http://dx.doi.org/10.1680/nectrpm.63532.029

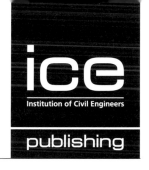

Starting on Site

This section describes how the *Project Manager* will ready themselves for work starting on the contract and describes what the *Project Manager* will need to do to manage the *works* in the way intended by the ECC.

5.1. Overview

It is important for the *Project Manager* to ensure that the contract processes are implemented from day 1. This can be achieved through setting up their documentation and systems to support the procedures required by the ECC. Failure to implement these at the outset will cause problems that could detract from the effective management of the contract.

Additionally, from day 1, the *Project Manager* should show leadership and commitment and ensure that they and their project team members follow the principles of clauses 10.1 and 10.2 and 'act as stated' and 'in a spirit of mutual trust and co-operation'.

To realise the benefits of the ECC, the key management processes, early warnings, programme, compensation events, payment and quality control among other items should be implemented from day 1 and maintained throughout the contract.

This chapter considers the key management processes and procedures that a *Project Manager* should ensure are in place at the commencement of a project.

Day 1 implementation

Start using the ECC procedures from the first day of the project.

5.2. Checklist

The *Project Manager* could have to hand the following agenda and checklist documents, to help them manage the project team and all the communications that take place between the project team. Some of these are items that are included in this book, and some can be found in other publications.

- Meeting agendas:
  - early meetings with other people working on the project (e.g. the *Supervisor*, any delegates, and perhaps even the *Client* if the *Project Manager* is required to report to a project sponsor)
  - kick-off meeting with the *Contractor*
  - weekly progress meetings on Site
  - monthly meetings with the *Contractor*
  - monthly meetings with other people in the project team
  - early warning meetings
  - quality management meetings (quality plan, Defects, tests/inspections and marking)
  - monthly payment meetings (main Options C, D, E and F).
- Communication templates, such as
  - early warning notification
  - *Project Manager's* instruction
  - compensation event notification
  - payment certificate.

- Pack of communication templates to be used by the *Contractor* and other project team members, such as
  - Defect notification
  - compensation event notification
  - *Supervisor's* instruction
  - test/inspection notification and result
  - early warning notification
  - *Contractor's* instruction to attend an early warning meeting
  - Defects Certificate
  - proposal for adding an area to the Working Areas.
- Wallchart with the outline programme, with Key Dates and key meetings marked.

5.3. The project team

The number of and types of people making up the project team will depend on the *Client's* approach to projects in general, the ECC in particular, and the main Option and programme under which the *works* are being built. (Figures 1.4, 1.6 and 1.7 in *NEC4: The Role of the* Supervisor (published by ICE, 2017) show some of the different project team configurations.)

The project team may comprise, for example,

- just the *Project Manager* (as they are also undertaking the role of *Supervisor*) and the *Contractor*
- a *Client's* representative, a project sponsor, a quantity surveyor (to whom the *Project Manager's* payment role has been delegated), a planner (to whom the *Project Manager's* programme duties have been delegated), a structural engineer and a design team reporting to the *Project Manager*, and three *Supervisors* (for the structural part of the *works*, the mechanical and electrical (M&E) systems part and the landscaping)
- anything in between these two scenarios.

Where the *Project Manager* has delegated, or has been required to delegate, some of their role to others, they will understand that they are still responsible for the whole of the role, and they will put in place procedures, systems and communications to make sure that they are able to undertake the overseeing role effectively. In particular, the *Project Manager* will make sure that the *Contractor* is aware of the delegations and who is the right person to communicate with on any issue, to ensure efficient working practices.

Two of the most important clauses in the ECC are clauses 10.1 (acting 'as stated in this contract') and 10.2 (acting 'in a spirit of mutual trust and co-operation'), and it is the *Project Manager's* duty to ensure that they and their delegates behave as required by and as implied in the ECC. Although there are some safeguards in the ECC to keep things moving forward in the absence of a required *Project Manager's* notification – deemed acceptance can take place as long as the *Contractor* makes the notifications they require (e.g. see clauses 31.3 and 61.4) – the *Project Manager* should try to make sure that there are no failures to communicate as required and that acceptances are provided as part of a conscious decision on their part and not because they did not act as stated.

Section 4.1 of this book provides more information about the project team set-up and delegation.

5.4. Reporting and management

The NEC suite of contracts is unique in the fact that if you follow the good management processes within a contract, then the regular reports should become a formalisation of the day-to-day management of these processes. NEC contracts are live contracts, and the data and information needed by the Parties should be a natural part of the project processes (Table 5.1).

The Scope should state the requirements for the timing, frequency and format of any regular reports required. The ECC Guidance Notes provide guidance on the types of information to be included in reports, and for this reason detailed examples are not included in this book.

Management meetings will also yield reports (Table 5.2). For example, the *Supervisor* will provide the *Project Manager* with a report on Defects prior to the monthly Defects meetings, and the *Contractor* will provide a report on progress for discussion at the monthly project meeting.

It is impossible for the *Project Manager* to be personally aware of everything that happens to the project and in the Working Areas, except on the smallest of projects, where the same person is undertaking both the *Project Manager* and

Table 5.1 Management procedures in the ECC

Procedure	Description
Early Warning Register	Updated for early warning notifications and at early warning meetings
Accepted Programme	Updated regularly, e.g. monthly
Compensation events	Change management events include assessment of changes to the Defined Cost, time and Key Dates
Defects and quality	Active management of the quality of the *works* through a quality management system, Defects notification and tests/inspections
Payment	Assessment of the amount due
Design management	The Scope will include design acceptance procedures

Table 5.2 Examples of regular meetings

Meeting	Reports produced for the meeting
Weekly progress meeting with the *Contractor*	▪ Programme update (the full Accepted Programme is not required)
Weekly meeting with the *Supervisor* about quality management issues	▪ Defects list ▪ Tests/inspections list and results
Monthly Defects meeting with the *Supervisor* and the *Contractor* (the *Contractor* can arrive after the first part of the meeting, if required)	▪ Defects list ▪ Proposals to accept Defects ▪ Access issues (if post-Completion) ▪ Defects that will affect Completion (towards the Completion Date) ▪ Proposals to reduce the number of Defects ▪ Issues with Subcontractors
Monthly early warning meeting with the *Contractor* and any other contributing people	▪ Early Warning Register ▪ List of early warnings ▪ Compensation events
Monthly progress meeting with the project team (including the *Contractor*)	▪ Progress ▪ Compensation events ▪ Prices/forecasts of total Defined Cost ▪ Plant and Materials
Monthly meeting with the *Client's* team (where delegation has taken place)	▪ Progress ▪ Budget ▪ Quality management

Supervisor roles. Even if they undertake the whole role of *Project Manager*, they will want to have regular meetings with the *Supervisor* and possibly the *Client*.

Appendices 8A and 8B contain sample agendas for weekly and monthly meetings.

5.5.   Ongoing design and design of Equipment
The procedures and duties of both the *Project Manager* and the *Contractor* in relation to design are very simple, but the two will need to set up the details early on so that the design takes place as smoothly as possible.

5.5.1   Design of the *works*
The thought process for the *Project Manager* is as follows.

1   Which parts of the *works* does the Scope require the *Contractor* to design?
2   What does the Scope require about how, when and in what way the *Contractor* should submit the design to the *Project Manager* for acceptance?

3   Can parts of the design be assessed fully without the other parts? The *Project Manager* should tell the *Contractor* because the decision to submit the design in parts is the *Contractor's* and not the *Project Manager's*, and there is no comeback for the *Project Manager* if the *Contractor* decides that each part can be assessed fully in isolation and so submits their design to the *Project Manager* for acceptance in parts.

4   The *Project Manager's* only reason for not accepting the *Contractor's* design is that it does not comply with either the Scope or the applicable law. Does the Scope provide any other reasons for non-acceptance?

5   The *Project Manager* must accept the design before the *Contractor* can proceed with the relevant work. What are the project impacts of delaying acceptance?

6   If the *Contractor* is required to obtain from a Subcontractor equivalent rights for the *Client* to use material prepared by the Subcontractor, how does the *Project Manager* want to check that this has been done?

If the design of the *works* is complicated and comprehensive, the *Project Manager* will want to set up progress meetings with the design team; procedures for checking on the design; and an efficient communication process, which allows questions and answers.

5.5.2   Design of Equipment

The thought process for the *Project Manager* is as follows:

1   Does the *Project Manager* want to see particulars of the design of an item of Equipment? If so, the *Project Manager* can instruct the *Contractor* to submit particulars of the design of an item of Equipment for acceptance.

2   The *Project Manager's* only reason for not accepting the *Contractor's* design is that it will not allow the *Contractor* to Provide the Works in accordance with the Scope, the *Contractor's* design that the *Project Manager* has accepted, or the applicable law. Does the Scope provide any other reasons for non-acceptance?

5.6.   Subcontracting
5.6.1   Subcontractors proposed with the tender

Most *Clients* will require tenderers to include with their tender a list of Subcontractors whom the tenderer intends to engage for the *works*. The invitation to tender might also contain questions on methods of procurement of the Subcontractors, and public sector bodies might also ask for the inclusion of certain terms and conditions, such as payment within 30 days of receipt of an invoice. The *Client* and its tender evaluation team have the opportunity to review the Subcontractors and ask questions, and they might also ask to see the subcontract documents (with or without pricing information) of the proposed Subcontractors before award is final so that, at award, both Parties know which Subcontractors have already been accepted at that stage.

The *Project Manager* can still use the procedure in clause 26.3 for reviewing the subcontract documents for the Subcontractors accepted at the tender stage if the review has not already been undertaken by the *Client* at the tender stage or the pre-award stage.

The procedure described in clause 26 is then reserved for the situation where, for example,

■   a Subcontractor who has been accepted can no longer provide their services to the *Contractor* and the *Contractor* is required to replace that Subcontractor or

■   through design development or changes in circumstance the *Contractor* now intends to subcontract other parts of the *works* for which they do not have Subcontractors.

5.6.2   Subcontract documents

The *Contractor* submits subcontract documents, excluding pricing information, for each subcontract to the *Project Manager* for acceptance unless

■   the proposed subcontract is an NEC contract that has not been amended, other than in accordance with the contract between the *Client* and the *Contractor*, or

■   the *Project Manager* has agreed that no submission is required

and the *Contractor* does not appoint a proposed Subcontractor until the *Project Manager* has accepted the Subcontractor and accepted the subcontract documents, including for main Options C to F, pricing information in the proposed subcontract documents.

In deciding whether to accept the proposed subcontract documents, the *Project Manager* will consider whether

- their use will allow the *Contractor* to Provide the Works
- they contain a statement that the parties to the subcontract act in a spirit of mutual trust and co-operation.

Options C to F require the *Contractor* to submit the pricing information in the proposed subcontract documents for each subcontract to the *Project Manager* unless the *Project Manager* has agreed that no submission is required. The purpose of the submission of the pricing information is not clear. Acceptance is not required. Perhaps the submission is just so that the *Project Manager* has a full picture of the *Contractor's* costs and is being provided with upfront evidence for section 4 of the SCC.

5.6.3 Subcontractors proposed post-award

If the *Contractor* intends to use a Subcontractor who was not proposed and accepted as part of the tender process, the *Contractor* must submit the proposed Subcontractor's name to the *Project Manager* for acceptance unless the *Project Manager* decides no submission is required. The *Project Manager* must accept or not accept within the *period for reply*. The only reason that the *Project Manager* may use for not accepting a proposed Subcontractor is that the appointment of the Subcontractor will not allow the *Contractor* to Provide the Works (and any other reasons included in the Scope).

There is no debate: if the *Project Manager* provides the non-acceptance with their reason, the *Contractor* may not appoint the proposed Subcontractor. Disagreements must go through the formal dispute route, but obviously the *Project Manager* will discuss their reasons with the *Contractor* (in the spirit of clause 10.2), and the two will agree a way forward.

5.7. Quality management

This book assumes that the *Supervisor* will keep site diaries that record progress and activities on Site and that the *Project Manager* can rely on these notes in their management of the project.

5.7.1 Relationship between the *Project Manager* and the *Supervisor*

The ECC places the monitoring and measurement of the quality of the *works* in the hands of the *Supervisor*, who has no direct relationship with the *Project Manager*. If the *Project Manager* wants to put in place procedures with the *Supervisor* to make sure that they are kept up to date on marking, tests/inspections and Defects, they will have to ask the *Client* to include those procedures in the *Supervisor's* personal contract or they will have to meet with the *Supervisor* before the *Contractor* starts on Site, and arrange meetings and documentation to suit.

The relationship between quality, time and budget may change if the following arrangements are put in place:

- In smaller projects, the *Project Manager* and the *Supervisor* might be the same person. In these sorts of projects, it is likely that the scope of the project is well defined and that the *works* will not take longer than a few months. Whatever decisions the *Project Manager* will be required to make can be made quickly and without consultation. The project can therefore progress more smoothly with one person than with two in the different roles. Tests/inspections and Defects should not occur frequently, and the *Project Manager* will be able to concentrate on their role rather than the *Supervisor's*.
- In other projects, the *Supervisor* might be delegated some of the *Project Manager's* duties, such as those in clauses 44.4, 45.1 and 45.2. There is no real conflict in this arrangement, as these sorts of duties fall into the quality management arena and are unlikely to impinge greatly on time and budget concerns.

5.7.2 General information about quality management

The *Supervisor's* role in quality management under the ECC has been extensively discussed in *NEC4: The Role of the Supervisor*, which describes various quality management systems.

The extent of the *Project Manager's* involvement in quality management depends on the type of project, the requirements of the *Client* and, to a lesser extent, the skills of the *Project Manager*. The ECC focuses on the *Project Manager's* duties in cost assessment, change assessment and access issues, and the *Client* will need to decide whether they also wish the *Project Manager* to carry out the communication duties in core clause section 4 or whether this will be required of the *Supervisor*. Appendix 9 summarises the *Project Manager's* and *Supervisor's* duties in managing the quality of the *works*. Tables 5.3, 5.4 and 5.5 describe the interactions between quality management and the *Project Manager's* cost assessments, change assessments and access duties, respectively. Table 5.6 lists communications about quality management.

Table 5.3 Interaction of the *Project Manager's* cost assessments with quality management

Clause	*Project Manager's* assessment	What the *Project Manager* wants the *Supervisor* to contribute
41.5	As part of their regular assessments of the amount due, the *Project Manager* will need to include in the relevant assessment the payment that is conditional upon a *Supervisor's* test/inspection being successful	■ Date of the test/inspection ■ Reason for the delay if the test has not been undertaken
41.6	Assesses the cost incurred by the *Client* of repeating a test/inspection after a Defect is found, e.g. ■ the cost of additional materials, samples and facilities ■ the cost of additional *Supervisor* time ■ the cost of Others or other people involved in the test/inspection (e.g. statutory performance)	■ What materials, samples and facilities were provided by the *Client* for the test/inspection ■ The *Supervisor's* role in the repeating of the test/inspection to calculate the cost of the *Supervisor's* involvement ■ Any independent contributors involved
C, D, E41.7	In an Option C, D or E contract, if a test/inspection needs to be repeated because the previous iteration revealed a Defect then, in assessing the cost of repeating the test/inspection, the *Project Manager* will consider only the *Client's* costs in carrying out the repeat test/inspection. In other words, the cost of the *Contractor* repeating the test/inspection is part of Defined Cost (unless it arises because the *Contractor* failed to follow a constraint stated in the Scope) and so the *Contractor* will be paid their costs of repeating the test/inspection, but they will have to pay the *Client's* costs of repeating the test/inspection (as assessed by the *Project Manager*)	
46.1	Assesses cost to the *Client* of having the Defect corrected by other people	■ Would a longer *defect correction period* allow the *Contractor* to correct the Defect? ■ Is the Defect correctable? ■ What materials, samples and facilities were provided by the *Client* for the test/inspection? ■ The *Supervisor's* role in the repeating of the test/inspection to calculate the cost of the *Supervisor's* involvement ■ Any independent contributors involved ■ What materials, samples and facilities and other things were provided by the *Contractor* for the test/inspection? ■ What is involved in the correction of the Defect – people, Equipment, Plant and Materials, and consumables? ■ How does the Defect affect what else is going on at the Site? ■ The approximate time that will be required to correct the Defect
46.2	Assesses cost to the *Contractor* of correcting the Defect	■ What materials, samples and facilities and other things would have been provided by the *Contractor* for the test/inspection? ■ What would have been involved in the correction of the Defect – people, Equipment, Plant and Materials, and consumables? ■ What extra costs would have been involved because of the location and difficulties now associated with the Defect (i.e. after take over)?

Table 5.4 Interaction of the *Project Manager's* change assessments with quality management

Clause	*Project Manager's* assessment	What the *Project Manager* wants the *Supervisor* to contribute
45.2	▪ Reduced Prices ▪ Earlier Completion Date	▪ What is the difference in the *works* with and without the Defect? ▪ Is there an impact on the performance of the *works*? ▪ What are the knock-on effects of the Defect? ▪ What cost factors would be incurred by the *Contractor* in correcting the Defect? ▪ How long would the Defect take to correct?
46.1	Change to the Scope; the Scope is treated as having been changed (no instruction required)	▪ What changes are required to the Scope?
46.2	Change to the Scope; the Scope is treated as having been changed (no instruction required)	▪ What changes are required to the Scope?

Table 5.5 Interaction of the *Project Manager's* access duties with quality management

Clause	*Project Manager's* duties	What the *Project Manager* wants the *Supervisor* to contribute
44.4	Arranges for access	▪ What access is required and to what part of the *works* to correct the Defect?
46.1	Was access provided?	▪ Was the right access given? ▪ Will longer access help?
46.2	Access is not given – advise *Supervisor*	No requirements

Table 5.6 Communications about quality management

Clause	Communication
41.2	If the Scope requires the *Client* to provide materials, facilities and samples for tests and inspections: ▪ The *Client* to advise the *Project Manager* how much notice they need to provide the materials, facilities and samples required by the Scope ▪ The *Project Manager* to ensure that the *Supervisor* notifies the *Project Manager* that the test/inspection will be carried out in X days (unless the *Project Manager* has it marked on the forecast weekly progress chart or it is a permanent item on the weekly progress meeting agenda) ▪ The *Project Manager* to provide appropriate notice to the *Client* ▪ The *Project Manager* to take delivery of items and liaise as appropriate to arrange for storage on Site of items if required ▪ The *Project Manager* to notify the *Supervisor* that the items are available to be used in the test/inspection ▪ The *Project Manager* to arrange disposal of items as required (unless the Scope requires the *Contractor* to dispose of them)
41.4	If a test/inspection is to be repeated: ▪ The *Project Manager* to ensure that the *Supervisor* advises the *Project Manager* that a test/inspection is to be repeated and to provide an estimated timeline ▪ Both the *Project Manager* and the *Supervisor* to undertake communications as for clause 41.2 ▪ The *Project Manager* to assess timelines and take action as required if the Accepted Programme will be affected

Clause	Communication
42.1	■ The *Supervisor* to notify the *Contractor* that Plant and Materials have passed their test/inspection if a test/inspection is required by the Scope ■ The *Project Manager* to ensure that the *Supervisor* notifies the *Project Manager* if the test/inspection has passed so that the *Project Manager* is aware of deliveries to the Working Areas ■ The *Project Manager* to ensure that the *Supervisor* notifies the *Project Manager* if a test/inspection was not passed, so that the *Project Manager* is aware of any potential delays (or the *Project Manager* can rely on an early warning notification from the *Contractor*)
43.1	■ The *Project Manager* to ensure that the *Supervisor* notifies the *Project Manager* that they have instructed a search, so that the *Project Manager* is notified of possible delays (or the *Project Manager* can rely on an early warning notification from the *Contractor*) and of the results of the search
43.2	■ The *Project Manager* to ensure that the *Supervisor* keeps the *Project Manager* notified of all Defects
44.2	■ The *Project Manager* to ensure that the *Supervisor* makes the *Project Manager* aware of when Defects might be corrected
44.4	■ The *Project Manager* to ensure that the *Supervisor* notifies the *Project Manager* that access is needed to correct a Defect – either at Completion when the correction of the Defect becomes due under clause 44.2 or at the time of notification of the Defect if the notification takes place after Completion ■ The *Project Manager* to advise the *Client* that access is required ■ The *Project Manager* to provide the *Contractor* with access as required
45.1	■ The *Project Manager* receives a proposal from the *Contractor* for acceptance of a Defect ■ The *Project Manager* seeks advice from the *Supervisor* about whether the proposal is viable and what the quality effects are

5.7.3 Approach to quality management

With the exception of accepting the quality policy statement and quality plan, it is the *Supervisor* who manages the quality management process; but some of the decisions that the *Project Manager* makes in their day-to-day management of the contract are affected by the results of tests/inspections, Defects or the status of Defects (e.g. deciding Completion), and the *Project Manager* will want to keep abreast of the quality management on the project. Tests/inspections and Defects can affect the project programme as well as the final price of the project, and the *Project Manager* can choose to be as involved as they want to be, sticking to their access, change and cost assessment duties, or liaising more closely with the *Supervisor* over the contract period.

Table 5.7 provides examples of the differing levels of involvement in the quality management process. Appendix 10 provides a *Project Manager's* checklist for managing quality, and Appendix 11 provides an agenda for a meeting about Defects and other quality matters.

It is clear that the *Project Manager* will find it difficult to undertake their duties as required by the ECC if they have no knowledge of how the quality management of the project is taking place on a day-to-day basis, and that they need reports from the *Supervisor* on tests/inspections and Defects during the period of the contract. However, the *Project Manager's* level of involvement will depend on the type of project and the personalities involved.

5.8. The Early Warning Register

Maintaining an Early Warning Register is good project management, and it can be used to keep track of those risks/issues that can affect the success of the project. The Early Warning Register in the ECC is defined in clause 11.2(8) and refers specifically to matters contained in the Contract Data and those notified as early warning matters. The Early Warning Register includes descriptions of the matters and the associated avoidance or reduction actions.

If the *Client* does not include a list of matters to be included in the Early Warning Register using the Contract Data, then it is the *Project Manager* who creates and then maintains the Early Warning Register, reviewing the matters included in CD2 by the *Contractor*. If the *Client* does not dictate how the Early Warning Register is to be maintained, then the *Project Manager* can choose how this is done, for example by using a standard impact/probability chart/matrix.

Table 5.7 Levels of *Project Manager* involvement in the quality management process

Project Manager does not want extra involvement	*Project Manager* wants regular updates of quality issues that may affect the Accepted Programme or amount due	*Project Manager* wants weekly/monthly updates
■ The *Project Manager* notes that the *Contractor* is operating a quality management system required by clause 40.1	■ The *Contractor* and *Supervisor* to provide confirmation to the *Project Manager* that the *Contractor* is operating a quality management system that complies with the requirements stated in the Scope	■ Both the *Supervisor* and the *Contractor* to provide confirmation at weekly progress meetings to the *Project Manager* that the *Contractor* is operating the quality management system as it was designed to be operated and that it complies with the requirements of the Scope; the *Supervisor* to log all non-conformances for reporting at the weekly meeting
■ The *Project Manager* accepts the *Contractor's* submitted quality policy statement and quality plan required to be submitted (clause 40.2)	■ The *Supervisor* to advise the *Project Manager* if the *Contractor* has changed the quality plan	■ The *Project Manager* to read weekly reports from the *Contractor* confirming that the latter has made no changes to the quality plan and is Providing the Works in accordance with its quality plan and quality policy statement
■ The *Project Manager* monitors the quality plan and instructs the *Contractor* to correct a failure to comply with the quality plan	■ The *Contractor* to self-report to the *Project Manager* any deviations from the quality plan that may affect the Affected Programme or the amount due	■ The *Contractor* to report weekly to the *Project Manager* that they have complied with the accepted quality plan
■ The *Project Manager* assesses the cost incurred by the *Client* in repeating a test/inspection after a Defect is found ■ The *Project Manager* arranges for the *Client* to allow the *Contractor* access to and use of a part of the *works* that they have taken over ■ The *Project Manager* receives proposals from the *Contractor* to accept a Defect and follows the rest of the procedure ■ The *Project Manager* may make proposals to the *Contractor* to accept a Defect ■ The *Project Manager* may accept a quotation to accept a Defect ■ The *Project Manager* assesses the cost to the *Client* of having the Defect corrected by other people under clause 46.1 and the cost to the *Contractor* of correcting a Defect under clause 46.2	■ For those tests/inspections that are on the critical path or that affect subsequent work, the *Supervisor* to notify the *Project Manager* when they are happening and their results ■ For regular tests/inspections, a statement of exception from the *Supervisor* (those that are not happening as planned or results are not positive) provided to the *Project Manager* ■ A list of notified Defects to date provided to the *Project Manager*, noting those that may affect the Accepted Programme or the amount due, those that have been corrected, and those that may prevent the *Client* from using the *works* or Others from doing their work ■ The results of all instructed searches provided to the *Project Manager* ■ Could any of the above lead to an early warning or a compensation event?	■ All notified Defects to date reported to the *Project Manager*, noting those that have been corrected and whether the *Supervisor* considers them to have been Defects that would have prevented Completion ■ Notification to the *Project Manager* of all forthcoming tests/inspections, so that the *Project Manager* can attend if required ■ Notification to the *Project Manager* of all tests/inspections that have taken place and their results

Matters to be added to the Early Warning Register are identified in two ways:

- those matters included in the Contract Data
- early warning matters notified by the *Contractor* or the *Project Manager*.

5.8.1   The Contract Data

Matters that are to be included in the Early Warning Register are included in both CD1 by the *Client* and CD2 by the *Contractor*, and are described as 'The following matters will be included in the Early Warning Register'. At this point it is likely that the matters will be described, but the manner of their avoidance and reduction will not, as this is not required by the Contract Data. The *Project Manager* should make sure that the matters listed in the Contract Data are discussed at the kick-off meeting and the manners of avoidance and reduction discussed and included in the Early Warning Register, after which the *Project Manager* should issue the revised Early Warning Register.

> - The following matters will be included in the Early Warning Register
>
>   ....
>   ....
>   ....

5.8.2   Early warning

Any early warning matters notified by the *Project Manager* or the *Contractor* will be entered by the *Project Manager* into the Early Warning Register. An early warning notification does not automatically lead to an early warning meeting to discuss the early warning, but the *Project Manager* and the *Contractor* are both allowed to instruct such a meeting. In many cases, an early warning meeting may not be required immediately if the early warning cannot be addressed straight away or if some investigation is needed to work out ways to mitigate the risk before the parties meet to discuss options, and these matters can be entered in the Early Warning Register and addressed at the next regular meeting. If an early warning meeting is to take place immediately, the agenda in Appendix 12B may be used to direct the meeting.

5.8.3   Early warning meetings

The first early warning meeting is held within e.g. 2 weeks of the *starting date*. The *Project Manager* instructs the *Contractor* to attend this first meeting. Regular meetings are then held until Completion of the whole of the *works*, and meetings can also be held if the *Contractor* or the *Project Manager* instructs the other to attend an early warning meeting.

Appendix 12A provides an agenda for a special early warning meeting, and Appendix 12B provides an agenda for a regular early warning meeting.

5.8.4   Maintaining the Early Warning Register

Clause 15.4 describes the *Project Manager's* requirement to revise the Early Warning Register at each early warning meeting and to issue it to the *Contractor* within 1 week after the meeting.

5.9.   The Accepted Programme

The Accepted Programme is arguably the most effective tool that the *Project Manager* has available to them. It has been described by some people as 'the beating heart' of the ECC. It should be kept up to date by the *Contractor*, as it forms one of the core documents and informs payment, compensation events and even Defects. The *Project Manager* needs to be proactive in working with the *Contractor* to ensure that there is always an up-to-date Accepted Programme.

The Accepted Programme is a joint management tool, and as such it is in both Parties' interests to ensure that there is an Accepted Programme in place. It is also worth noting that there is no 'I accept but ...' in the ECC: there is only acceptance or non-acceptance.

Keeping the Accepted Programme current is time consuming and requires fairly intensive management. The *Project Manager* may therefore choose to delegate to a planner/programmer the actions related to the programme.

The *Project Manager's* actions regarding the Accepted Programme are

- receiving the *Contractor's* programme, which is submitted at the intervals stated in the Contract Data
- responding to the *Contractor's* programme – either acceptance or an instruction to revise the programme.

The *Project Manager* should also check for other items that will affect the Accepted Programme, such as

- early warnings discussed at an early warning meeting that are accompanied by a programme
- quotations for compensation events that include a programme
- definition of Price for Work Done to Date – work that contains no Defects that will be covered by follow-on activities.

5.9.1   FAQs about the Accepted Programme

Table 5.8 provides responses to some frequently asked questions about the Accepted Programme.

Table 5.8 FAQs about the Accepted Programme

What is the Accepted Programme?	It is either (a) the programme submitted at the time of tender by the *Contractor* with their tender or (b) a programme submitted within X weeks of the Contract Date by the *Contractor*. When the *Project Manager* accepts it, it becomes the Accepted Programme.
Is there only one Accepted Programme?	There is only one Accepted Programme at any given time. Programmes that were an Accepted Programme but have now been superseded are not referred to in the ECC. (For the purposes of this book, these are called superseded programmes.)
When does the *Contractor's* programme become the Accepted Programme?	Each programme submitted by the *Contractor* to the *Project Manager* is simply called the programme, and it is only when it is accepted by the *Project Manager* that it becomes the Accepted Programme.
When does the *Contractor* submit a programme to the *Project Manager*?	The *Contractor* can submit a programme to the *Project Manager* at any time (clause 32.2), and they are required to submit one at the beginning of the contract (at tender stage or a number of weeks after the Contract Date), when they are instructed to and regularly through the contract period (clause 32.2).
Can the *Project Manager* not accept the *Contractor's* programme?	The simple answer is yes they can. Clause 13.4 makes it clear that if the *Project Manager's* reply is not acceptance, the *Project Manager* states their reasons.
	Clause 31.3 lists four reasons why the *Project Manager* may choose to not accept a revised programme submitted for acceptance by the *Contractor*:
	▪ It shows the *Contractor's* plans are not practicable, e.g. the programme does not allow for the time to get security clearance or passes for their people to work on Site.
	▪ It does not show the information that the contract requires, e.g. clause 31.2 and the last bullet point of clause 31.2 'other information that the Scope requires the *Contractor* to show on a programme submitted for acceptance'.
	▪ It does not show the *Contractor's* plans realistically, e.g. the output rates, time durations and resources are not achievable.
	▪ It does not comply with the Scope, e.g. it is not provided in the format or software package required by the Scope.
	Clause 13.8 makes it clear that withholding acceptance of a submission by the *Contractor* for a reason stated in the contract is not a compensation event. If the *Project Manager* withholds an acceptance for a reason other than the four reasons stated in clause 31.3 (and the generic reason in clause 13.4), then this would in principle be a compensation event under clause 60.1(9). The *Contractor* would still have to demonstrate that the withheld acceptance met the tests as stated in clause 61.4, e.g. it has an effect on the Defined Cost, Completion or meeting a Key Date.
What if the *Project Manager* does not accept the *Contractor's* programme but does not instruct the *Contractor* to submit a revised programme?	▪ The current Accepted Programme remains the Accepted Programme and the *Contractor* Provides the Works in accordance with that Accepted Programme. ▪ In accordance with catch-all clause 13.4, the *Contractor* must resubmit the programme within the *period for reply*.

What if the *Contractor* doesn't agree with the *Project Manager's* reasons for non-acceptance?	■ The current Accepted Programme remains the Accepted Programme and the *Contractor* Provides the Works in accordance with that Accepted Programme. ■ The *Contractor* must still submit a revised programme under clause 13.4 or, if the *Project Manager* has instructed them to resubmit, clause 32.2. ■ The *Contractor* can meet with the *Project Manager* and request further information to understand better what the *Project Manager* expects to see. ■ The *Contractor* can refer a dispute in accordance with Option W1, W2 or W3.
What if the *Contractor* has not submitted a programme?	■ If the *Contractor* was supposed to have submitted a first programme (either with the tender or after contract award) and did not do so, then clause 50.5 allows the *Project Manager* to retain 25% of the Price for Work Done to Date from the amount due assessed in each assessment until the *Contractor* submits the first programme to the *Project Manager* for acceptance and it includes the information that the contract requires to be included. ■ As the programme is a key part of the project, it seems unlikely that the *Project Manager* will not have instructed the *Contractor* to submit a first programme at some point in the weeks before the first assessment date, although clause 50.5 provides a useful sanction. The other sanction on the *Contractor* is that if there is no Accepted Programme then the *Project Manager* makes their own assessment of compensation events. A more likely scenario is that the *Project Manager* will sit down with the *Contractor* and try to find out what the underlying cause is so that they can both work together to get an Accepted Programme in place.
What if the *Project Manager* does not reply to the *Contractor* about their submitted programme?	■ The current Accepted Programme remains the Accepted Programme and the *Contractor* Provides the Works in accordance with that Accepted Programme (and not the recently submitted programme that has not been accepted). ■ Clause 31.3 provides that the *Contractor* can notify the *Project Manager* that the *Project Manager* has not notified acceptance or non-acceptance within the time allowed (2 weeks after submission of the programme to the *Project Manager* or a longer period that has been agreed between the *Contractor* and the *Project Manager*). The *Project Manager* then has a further week to notify acceptance or non-acceptance of the programme to the *Contractor*. ■ If the *Contractor* does not receive notification of acceptance or non-acceptance by the *Project Manager* by the end of that final week, then the *Contractor* may assume that the *Project Manager* has accepted the programme and that the submitted programme is now the Accepted Programme. ■ The *Contractor* may choose to notify the *Project Manager* that deemed acceptance has taken place.
What if the *Project Manager* takes longer to reply than 2 weeks (as required by clause 31.3)?	If the *Project Manager* takes longer than 2 weeks to reply, then this is a compensation event under clause 60.1(6). If the *Contractor* can demonstrate that the lack of a response has affected their Prices, Completion or a Key Date, then the *Contractor* will be entitled to compensation as described in clause 63.

5.9.2 How the programme is used in other procedures

Not only does the Accepted Programme interact with the other key project documents, it also has a relationship with some of the procedures in the ECC. The early warning and compensation event procedures are two of the key procedures in which the Accepted Programme plays a part.

5.9.2.1 The Accepted Programme and early warnings

Early warnings are one of the ways in which the *Project Manager* keeps track of what is happening on the Site and how the *works* are progressing. Two of the reasons for notifying an early warning affect the Accepted Programme are

■ a matter that could delay Completion
■ a matter that could delay meeting a Key Date.

The *Project Manager* will be cognisant that a matter that could increase the total of the Prices or impair the performance of the *works* in use could also impact on the Accepted Programme. The *Project Manager* is therefore likely to have the Accepted Programme to hand during early warning meetings, when discussing proposals for mitigation and solutions to the early warning matter.

> Although clause 15 refers to the *Project Manager* and the *Contractor* notifying each other of an early warning, the *Project Manager* may also receive relevant information from other sources, such as the *Supervisor*.

Part of any solution to an early warning will involve reviewing the Accepted Programme and discussing how it will be impacted by proposed solutions so that those attending the early warning meeting can decide on the most effective solution as a combination of the programme, costs and quality/performance.

The *Supervisor* can act as the *Project Manager's* eyes and ears on the Site, and as such they can provide invaluable insight into the progress of activities on Site.

The *Project Manager* should hold regular meetings with the *Supervisor* and consult with them on issues in relation to compensation events, Defined Cost, completed activities and so on.

5.9.2.2 The Accepted Programme and compensation events

The Accepted Programme provides inputs to the compensation event procedure. It also forms a part of the output of the procedure.

Compensation events are affected by the Accepted Programme as follows:

- clause 60.1(2) – provide access by the later of its *access date* and the date shown on the Accepted Programme
- clause 60.1(3) – provide something by the date shown on the Accepted Programme
- clause 60.1(4) – instruction to change a Key Date
- clause 60.1(5) – work within the times shown on the Accepted Programme
- clause 60.1(11) – test/inspection causes unnecessary delay
- clause 60.1(15) – take over before the Completion Date
- clause 60.1(16) – does not provide things for tests/inspections
- clause 60.1(19) – stops Completion being achieved by the date shown on the Accepted Programme.

Equally, the Accepted Programme can be affected by compensation events. Quotations for compensation events provided by the *Contractor* include proposed delays to the Key Dates and the Completion Date assessed by the *Contractor*. Delays to the Completion Date are shown on the alterations to the Accepted Programme in relation to the date for planned Completion. The implementation of the compensation event means that the assessment/quotation for the compensation event is accepted, and the quotation includes changes to the Accepted Programme, so that every programme submitted after the implementation of the compensation event includes the impact of the compensation event.

5.9.2.3 The Accepted Programme and payment

The Accepted Programme is linked with payment because the *Contractor* is paid as the work progresses. In addition, the *Contractor* will programme the work so that their cash flow is maximised. There is a more obvious link in Option A and C contracts, as the *Contractor* needs to show how the activities in the Activity Schedule relate to the programme.

The *Project Manager* will use the Accepted Programme in their assessment of the amount due, and will consider

- which activities on the Activity Schedule are complete (Option A)
- what quantity of work has been completed for each item in the Bill of Quantities (Option B)
- what is forecast to have been paid by the next assessment date; that is, what operations are shown on the Accepted Programme as being undertaken next (Options C, D, E and F).

The *Project Manager* is likely to have the Accepted Programme next to them when they are assessing the amount due, so that they can review what work might be covered by the assessment: completed work for Options A and B; and forecast work for Options C, D, E and F.

Section 6

NEC4: The Role of the *Project Manager*
ISBN 978-0-7277-6353-2

Chronological management procedures: first month/period

This chapter discusses the procedures and actions that affect the *Project Manager* during the first month after the *starting date*.

6.1. Regular meetings

NEC contracts are all about good management, and as such the *Project Manager* should establish working practices which enable the key NEC processes to succeed. Again, as the NEC is a 'live' contract, any regular meetings (e.g. weekly/monthly) should be a formalisation of the everyday management processes (management and reports are discussed in Section 5.4 of this book).

Regular meetings that involve the *Project Manager* may include

- weekly progress meeting with the *Contractor*
- weekly meeting with the *Supervisor* about quality management issues
- monthly Defects meeting with the *Supervisor* and the *Contractor* (the *Contractor* can arrive after the first part of the meeting if required)
- monthly risk reduction meeting with the *Contractor* and any other contributing people
- monthly progress meeting with the project team (including the *Contractor*)
- monthly meeting with the *Client's* team (where delegation has taken place).

6.2. First programme
6.2.1 When is a first programme submitted?

The *Contractor* is required to submit a first programme to the *Project Manager* for acceptance either at the tender stage or within a number of weeks after the Contract Date, depending on whether the *Client* wants to see the programme at the tender stage or whether they are content to wait until after contract start to see the programme. The *Client's* choice will be based on the type, length and complexity of the project, and the *Project Manager* is unlikely to have the opportunity to influence their decision.

6.2.1.1 First programme submitted with the tender

If the *Contractor* is required to submit or has submitted a first programme as part of their tender, CD1 by the *Client* will not mention the first programme, and CD2 by the *Contractor* will have the following entry, which will have been completed by the *Contractor* as part of their tender.

CD2 – data provided by the *Contractor*

The programme identified in the Contract Data is . . . . [e.g. the document labelled Programme v. 01]

6.2.1.2 First programme submitted after contract award

If the *Contractor* is **not** required to submit a first programme as part of their tender, but rather at a point in time after the contract has started, then CD2 (by the *Contractor*) will not mention the programme, and CD1 will have the following entry, which will have been completed by the *Client* in the invitation to tender with a specified number of weeks.

CD1 – data provided by the *Client*

■ The period after the Contract Date within which the *Contractor* is to submit a first programme for acceptance is . . . . [e.g. 4 weeks]

It is worth mentioning that, since the *completion date* is tied to the programme, a *Client* is likely to have provided the *completion date* in CD1 so that, even though they have not evaluated the whole programme, they have taken steps to require the *Contractor* to complete the project within the *Client's* time-frame. It is also possible that the *Client* will have requested an outline programme (this is not ECC terminology) at the tender stage, so that the *Contractor's* proposal can be evaluated in competition with other tenderers' proposals, and it wants the full programme after contract award, in line with clause 31.2.

6.2.1.3 Implications for the *Project Manager*

It may be that the *Project Manager* is not involved in the project at the tender stage and so will not see a first programme until they start their job as the *Project Manager*, which may well be after contract award. When the *Project Manager* starts working on the project, they might have access to the programme as one of the following scenarios:

■ **No programme**: the *Client* did not want to see a programme at the tender stage, and CD1 asks for a programme to be provided a number of weeks after the Contract Date (clause 11.2(4) – the Contract Date is when the contract comes into existence) because it has provided a *completion date* in CD1 and so is satisfied that the *Contractor* is required to reach Completion before the *completion date*. Since the *Project Manager* starts their job after contract award and there was no programme at the tender stage, the *Project Manager* will not have access to a programme.

■ **Outline programme**: the *Client* did not want to see a full programme at the tender stage, but it did want to evaluate the *Contractor's* approach to the programme and so requested an outline programme in the invitation to tender. This outline programme would be available to the *Project Manager* when they come on board. This approach is useful if one of main Options C to F is being used. Receipt of a full programme is more suitable for Option A and B contracts, but the *Client* may still ask to see just an outline programme at this stage, especially if the project is being tendered among several contractors.

■ **Fully detailed programme**: the *Client* requested a full programme at the tender stage by including the relevant entry in CD2, and the *Contractor* submitted a programme with their tender.

6.2.2 Sources of information in a first programme

Appendix 13 lists the source documents where the *Project Manager* will find the information that the *Contractor* must include in their first programme submitted for acceptance.

6.2.3 Information to be included in a first programme

Appendix 14 provides a checklist for the dates, points in time and information that the *Contractor* must include in their first programme submitted for acceptance.

Note that this checklist lists the programme elements required by the ECC; it is not a work instruction describing what the *Project Manager* should look for when reviewing a programme submitted for acceptance by the *Contractor*.

6.3. First assessment of the amount due

The *Project Manager* will want to set up their procedures so that everything is ready before the first assessment takes place. For Option C, D or E contracts, the *Project Manager* is likely to engage the *Contractor* in discussions, perhaps as part of the kick-off meeting, to make the *Contractor* aware of the Scope requirements and to find out about the *Contractor's* expectations of the assessment. The *Project Manager* may want to discuss the form of the *Contractor's* application for payment detailed in the Scope and other details that could make the review of the application easier.

If secondary Option Y(UK)2 is used in the contract, the *Project Manager* will include those extra parts of the procedure into their monthly schedule.

6.3.1 First assessment date: when does it occur?

The *Project Manager* is required to decide when the first assessment date occurs. In doing so, they are required to take into account the procedures of the Parties, and the date cannot be later than the *assessment interval* after the *starting date* (clause 50.1).

6.3.2   What the *Project Manager* needs to know in order to decide the first assessment date

The *Project Manager* needs to have the following information:

- The *starting date* – found in CD1 under section 3.
- The *assessment interval* – found in CD1 under section 5. This is usually expressed as a number of weeks, but it cannot be more than 5 weeks. Alternatively, the Contract Data may provide a schedule of assessment dates or it may describe an interval of time, such as the time period between the previous assessment date and the last Thursday of each calendar month.
- The Accepted Programme – the *Project Manager* may want to make the first assessment date after the first programme is due. The date for the submission of the first programme will be stated in CD1.
- Procedures of the Parties – found in the Scope and tender documents and in informal/verbal communications. At this stage there is only one procedure that affects the time between the *starting date* and the first assessment date:
  - secondary Option X14 (advanced payment to the *Contractor*): the *Client* requires one payment to be made that includes both the advanced payment and the first assessment of the amount due.

6.3.3   What information the *Project Manager* needs to collect before undertaking their first assessment of the amount due

Appendix 15 lists the information that the *Project Manager* needs to collect before undertaking their first assessment of the amount due.

6.3.4   How the first assessment of the amount due takes place

Exactly how the first assessment is made will depend on which main Option is being used – the main Option determines the method of payment. There are some common elements in the methods of assessment across the various main Options, but, as is to be expected, there are some slight differences between them.

6.3.4.1 Option A

The *Project Manager* should check whether the Activity Schedule describes sufficient activities so that the *Contractor* is assured of some income for each month of the project. This is not part of the *Project Manager's* role and it is not their responsibility to ensure the *Contractor's* cash flow; but if the *Project Manager* notices that the Activity Schedule put forward by the *Contractor* may not assure cash flow as the project progresses, they may wish to point this out to the *Contractor* to give them the option of amending the Activity Schedule by breaking up activities into smaller parts. Since the Activity Schedule is reflected in the *Contractor's* programme, any change to the Activity Schedule should be reflected in the programme as well – the *Contractor* can revise the programme when they choose to.

The *Contractor* takes the risk of putting together an Activity Schedule subject to any requirements stated in the invitation to tender.

In assessing the amount due, the *Project Manager* considers the activities described in the Activity Schedule and assesses whether the activities are complete or not. If activities are in a group, then all the activities must be complete for the group to be assessed as complete.

The *Project Manager* will consider the following:

- What is the description of the activity in the Activity Schedule? It is helpful if the activities are described as being complete: for example, rather than describing an activity as 'The installation of M&E systems', it may be more helpful to describe it as 'The staff room has fully functional electrical systems'.
- Are there any Defects (notified or not) that would (1) delay immediately following work or (2) be covered by immediately following work? This determination can be made through comparison with the Scope, and therefore it would be helpful if the Scope includes appropriate descriptions of activities. This may not be relevant where the *Contractor* decided the activities included in the Activity Schedule. The *Project Manager* may want to engage with the *Supervisor* before assessments, to ask to be alerted where Defects may affect an assessment of activity completion.
- Does the Scope describe what is meant by a complete activity?

The Accepted Programme is crucial if the assessment of the amount due is to be accurate, and therefore the *Project Manager* can retain one-quarter of the PWDD in amounts due if the *Contractor* has not submitted a programme for acceptance by the first assessment date (clause 50.5) (assuming that CD2 does not provide for a programme to be submitted with the tender and that CD1 does not require a programme to be submitted a number of weeks after the contract start that takes the first submission past the first assessment date). Note, too, that the clause refers to the **submission** of the programme, rather than to its acceptance by the *Project Manager*; this gives the *Contractor* some comfort if the *Project Manager* is tardy in their review of the programme.

The process for compensation events concludes with changes to the Prices, the Accepted Programme and Key Dates being implemented (clause 66.1). Therefore, at the time of assessing the amount due, the compensation event is either complete or not: there is no in-between, where a part of the compensation event may be included in an assessment of the amount due. If the process has been completed, the compensation event will have changed the Prices and so any assessment of activities takes into account the changed Prices. If the compensation event process is not yet complete, the Activity Schedule will reflect the pre-compensation event situation, and the assessment must take place without taking into account the changed position.

The *Project Manager* and *Contractor* should take care in implementing compensation events, as once implemented they cannot be revised (clause 66.3).

6.3.4.2 Option B

As with Option A, the regular assessments of the amount due are based on the pricing document submitted with the tender – in this case, the Bill of Quantities (the use of the SSCC is limited to providing quotations for compensation events). The *Client* prepares and takes the risk for any errors in the Bill of Quantities.

In assessing the amount due, the *Project Manager* will take into account the rates in the Bill of Quantities against the quantities of items that are completed. The nature of the Bill of Quantities means that cash flow is more readily assured for the *Contractor*, but the *Project Manager* will naturally spend more time in making the assessment, as they will be required to assess the quantities for each item in the Bill of Quantities.

The *Project Manager* will take the following into account:

- What items are listed in the Bill of Quantities and what quantities of them have been used during the period from the *starting date* to the first assessment date?
- Are there any Defects (notified or not) that would (1) delay immediately following work or (2) be covered by immediately following work? This determination can be made through comparison with the Scope, and therefore it would be helpful if the Scope includes appropriate descriptions of any completed parts of the *works* that can be isolated on the Accepted Programme.

The Accepted Programme is crucial if the assessment of the amount due is to be accurate, and therefore the *Project Manager* can retain one-quarter of the PWDD of amounts due if the *Contractor* has not submitted a programme for acceptance by the first assessment date (clause 50.5) (assuming that CD2 does not provide for a programme to be submitted with the tender and that CD1 does not require a programme to be submitted a number of weeks after the contract start that takes the first submission past the first assessment date). Note, too, that the clause refers to the **submission** of the programme rather than its acceptance by the *Project Manager*; this gives the *Contractor* some comfort if the *Project Manager* is tardy in their review of the programme.

The process for compensation events concludes with changes to the Prices and the Accepted Programme and Key Dates being implemented (clause 66.1). Therefore, at the time of assessing the amount due, the compensation event is either complete or not, and the Bill of Quantities will reflect the change. The compensation event process has the potential to be quicker than in other main Options because assessments for changed Prices for compensation events use the rates and quantities in the Bill of Quantities and the *method of measurement* (clause 63.15) rather than the SSCC.

6.3.4.3 Options C and D

Regular assessments of the amount due in Option C and D contracts are time consuming and challenging and can be very difficult if the *Contractor* and the *Project Manager* are not working together. If a *Client* has chosen Option C or D as the method of payment because it cannot spare the time to provide more certainty in the Scope, it must then understand that their *Project Manager* will spend a large part of their time on assessing the amount due.

The Price for Work Done to Date is not directly related to the Activity Schedule/Bill of Quantities provided by the *Contractor* as part of their tender; instead, it requires constant reference to CD2 by the *Contractor* for its *fee percentage* and data for the SCC. The *Project Manager* must also refer to evidence of payments and charges as described in the SCC and provided by the *Contractor*, and they may want to consider setting up a process that electronically records references to documents to facilitate a permanent and contemporaneous library of payments throughout the project.

For cost-based contracts using Options C, D or E, the *Contractor's* application for payment will help to speed up the *Project Manager's* assessment. It can also help prevent disputes, and ease conversations about the amount due.

6.3.4.4 Option E

As there is no initial pricing document used in Option E and no tendered total of the Prices, there is no comparison to be made between a price that was intended and the actual prices calculated as the project goes on. However, there is likely to have been some kind of project document that can be used as a base for project cost; the *Project Manager* may be able to use this to increase understanding of the Price for Work Done to Date and whether the cost of the project is on track and reasonable.

An Option E contract can start with very little research and procurement time, and so can be fast moving as far as the assessment of the amount due is concerned, with the potential for myriad compensation events. The *Project Manager* will need to be alert as far as the inclusions for Defined Cost are concerned and check that the *Contractor* is applying Defined Cost correctly (e.g. they are excluding items of Disallowed Cost). The setting up of the documentation and software and the rigour of the assessment procedure will be crucial to the *Project Manager's* successful assessments.

6.3.4.5 Option F

The SCC is not used in Option F contracts, and therefore much of the amount due will be based on what the *Contractor* has paid their Subcontractors, with the *Contractor's* Prices being lump sum amounts as stated in CD2. Compensation events can result in changes to the Prices, but the focus is on payments to Subcontractors and the Prices for work done by the *Contractor* themselves. Any compensation event in relation to work the *Contractor* is to do themselves is subject to agreement between the *Contractor* and the *Project Manager*. If they cannot agree, the *Project Manager* decides the change.

6.3.4.6 First assessment checklist for all main Options

Appendix 16 contains a checklist for the first assessment of the amount due (all main Options). An example payment certificate is included in the ECC Guidance Notes.

6.4. First early warning notification

Early warnings can occur throughout the project, but if the procedure is set up at the beginning, then it creates the right expectations and facilitates a smoother process as the contract goes on. All those involved will be aware of their responsibilities and how their actions and decisions affect others in the project.

> Early warnings help to mitigate problems and shrink risk. They help to provide greater certainty of outcome in terms of the Prices, the Accepted Programme and the performance of the *works*.

Table 6.1 lists the clause requirements of the early warning procedure, and Table 6.2 describes the implications of setting up the early warning procedure. Appendix 17 provides a checklist for the notification of an early warning.

6.4.1 Consequences

Clauses 61.5 and 63.7 describe the consequences of a *Contractor* not providing an early warning, which are summarised in Figure 6.1.

6.5. First compensation event

A compensation event is not a defined term in the ECC, but it can be described as an event that, if it were to occur, may entitle the *Contractor* to more time or money.

There are four sources of compensation events:

- clause 60.1 (21 reasons – the last of which refers to additional compensation events included in CD1)
- main Option B and D clauses 60.4, 60.5 and 60.6 (a further three reasons)
- secondary Option clauses X2.1, X12.3(6) and (7), X14.2, X15.2 and Y2.5
- potentially, secondary Option Z (which could list additional compensation events).

Note: clause 60.1(14) refers to an event that is a *Client's* liability stated in this contract. This takes you to clause 80.1; the final bullet point of this clause refers to additional *Client's* liabilities stated in the Contract Data.

Table 6.3 lists compensation events and their links to other clauses in the contract.

Table 6.1 Clause requirements of the early warning procedure

Originating from the *Project Manager*	Originating from the *Contractor*
■ The *Project Manager* can notify the *Contractor* of an early warning.	■ The *Project Manager* can receive a notification of an early warning from the *Contractor*.

■ As a 'notification', the early warning must be
  – in writing ('read, copied and recorded') (clause 13.1)
  – communicated through the communication specified in the Scope or received at the address in the Contract Data or later notified (clause 13.2)
  – communicated separately from other communications (clause 13.7).
■ The notification must be made 'as soon as' the notifier becomes aware of the matter (clause 15.1).	

■ The reasons for notifying an early warning are that the matter could
  – increase the total of the Prices
  – delay Completion
  – delay meeting a Key Date
  – impair the performance of the *works* in use
  – be any other matter that could increase the *Contractor's* total cost.
	■ The reasons for notifying an early warning are that the matter could
	  – increase the total of the Prices
	  – delay Completion
	  – delay meeting a Key Date
	  – impair the performance of the *works* in use
	  – be any other matter that could increase the *Contractor's* total cost.

■ The *Project Manager* can instruct the *Contractor* and can also instruct other people to attend an early warning meeting (if the *Contractor* agrees). The *Contractor* is required to obey an instruction given to it by the *Project Manager* and which is in accordance with the contract and so must attend the meeting.	■ The *Project Manager* can be instructed by the *Contractor* to attend an early warning meeting. The *Contractor* can also instruct other people to attend the same risk reduction meeting if the *Project Manager* agrees.

Table 6.2 Implications of setting up the early warning procedure

Clause requirement	Implication for the *Project Manager*
The *Contractor* can instruct the *Project Manager* and also other people to attend the early warning meeting.	The *Project Manager* should be prepared to be in a meeting where they are unaware of all the other attendees. The *Project Manager* may want to put in place some procedures for the *Contractor* to follow if other people are going to attend the early warning meeting.
The *Project Manager* and the *Contractor* give notification.	The *Project Manager* may feel that it is useful for the *Supervisor* and/or other key project people (e.g. a quantity surveyor, if delegated *Project Manager* payment duties, and even the *Client*) to have the right to provide early warning notification under the contract. The *Project Manager* may want to put in place procedures to facilitate this, perhaps encouraging discussion at weekly meetings prior to formal notification of an early warning.
Notification must take place as soon as the *Project Manager* or the *Contractor* 'becomes aware' of the matter.	Becoming aware of the matter is different from becoming aware that the matter could affect the Prices, programme and performance of the *works*. The *Project Manager* may want to discuss the parameters of this assessment so that all those involved in the project are aware of what the *Project Manager* is measuring as part of the procedure.
The matter could affect the Prices, the programme and the performance of the *works*.	The *Project Manager* will need to make sure that they have access to the Accepted Programme and will ensure that the Accepted Programme is revised as required by the project (which may require more or less frequent revision than is required by the ECC). The *Project Manager* will also make sure that the Prices are kept current, that they have an idea of the *Contractor's* total Defined Cost for the whole of the *works* and that they have comprehensive knowledge of the Scope so that they understand whether an early warning matter will affect the performance on the *works*.

Figure 6.1 Consequences of the *Contractor's* failure to provide an early warning

The *Project Manager* can decide that the *Contractor* did not give an early warning that an experienced contractor could have given, and can notify this to the *Contractor* when they instruct the *Contractor* to submit quotations for a compensation event

The compensation event is assessed as if the *Contractor* had given an early warning

Therefore, no matter who assesses the compensation event, the *Project Manager's* decision could affect the options available to the *Contractor* in addressing the compensation event

The compensation event process provides timelines for communications and replies, rather than using the standard *period for reply*; the *Project Manager* and the *Contractor* are obliged to adhere to these timelines (clause 13.3). However, the *Project Manager* may extend the *period for reply* for any of the described communications if the *Project Manager* and *Contractor* agree before the reply is due (clauses 62.5 and 13.5).

Table 6.3 Linking compensation events to other clauses

Clause	Description	Relevant clause
60.1(1)	Change to the Scope	14.3, 27.3, 45
60.1(2)	Access to and use of the Site	33.1 and 31.2
60.1(3)	*Client* providing something	31.2
60.1(4)	Stop or not start any work	34.1
60.1(5)	*Client's* and Others' working times and conditions	31.2
60.1(6)	Replying to communications	13.3
60.1(7)	Object of value	73.1
60.1(8)	Changing decisions	–
60.1(9)	Withholding acceptance	13.8, 13.4, 24.1, 31.3, etc.
60.1(10)	Instructions to search	43.1
60.1(11)	Test or inspection causing delay	41.5
60.1(12)	Physical conditions	60.2, 60.3
60.1(13)	Weather	–
60.1(14)	*Client's liability*	80.1
60.1(15)	Take over	35.2
60.1(16)	*Client* provides materials, facilities and samples	41.2
60.1(17)	Correction to an assumption	61.6
60.1(18)	Breach of contract	–
60.1(19)	Unforeseen events	19.1
60.1(20)	Quotation for a proposed instruction	65.3
60.1(21)	Additional compensation events	CD1

Table 6.4 The five key stages of the compensation event procedure

Stage	Actions	Timescale
1. Notify	Is the event notified as a compensation event in principle?	
	The *Project Manager* is required to notify compensation events under clauses 60.1(1), (4), (7), (8), (10), (15), (17) and (20).	Notification by the *Project Manager* takes place at the same time as the event.
	The *Contractor* is required to notify compensation events under clauses 60.1(2), (3), (5), (6), (9), (11), (12), (13), (14), (16), (18), (19) and (21).	Notification by the *Contractor* takes place within 8 weeks of the event. These compensation events can be time barred by virtue of clause 61.3 – entitlement will be lost if the *Contractor* does not notify within 8 weeks of becoming aware of the matter.
2. Instruct	The *Project Manager* instructs the *Contractor* to submit quotations.	At the same time as notification and the event.
	The *Project Manager* notifies the *Contractor* of their decision and instructs the *Contractor* to submit quotations.	Within 1 week following the *Contractor's* notification. (If the *Project Manager* does not do this by 2 weeks after the *Contractor's* notification, an instruction to submit quotations is assumed.)
3. Quote	The *Contractor* prepares the quotation after discussing different ways of dealing with the compensation event.	The *Contractor* has 3 weeks to submit the quotation.
4. Assess	The *Project Manager* should consider the sources of information and the records and facts required to enable the assessment of compensation events.	The *Project Manager* has 2 weeks to assess the compensation event.
5. Implement	■ The *Project Manager* notifies acceptance of the *Contractor's* notification or ■ the *Project Manager* notifies the *Contractor* of their own assessment or ■ the *Contractor's* notification is treated as having been accepted by the *Project Manager*.	■ The *period for reply*. ■ 3 weeks. ■ More than 3 weeks.

The compensation event procedure has five key stages, which are laid out in Table 6.4.

There are obvious links between early warnings and compensation events. Apart from those shown in Section 6.4.1 of this book, it makes sense for the *Project Manager* to keep an eye on the Early Warning Register and compensation events – in general, more issues could be raised and resolved in an effective and risk-reducing manner if they were raised as early warnings rather than when they become compensation events.

Both procedures require quick action: early warnings are required to be notified as soon as the notifier becomes aware of a relevant matter; and compensation events are required to be notified by the *Contractor* as soon as they become aware of one occurring, with an 8-week cut-off period for certain events.

While the compensation event procedure requires actions and communications from both the *Contractor* and the *Project Manager*, the *Project Manager* could lead the procedure and make sure that the *Contractor* does what they are required to do so that the procedure ends quickly and effectively.

Table 6.5 lists decisions that the *Project Manager* is required to make during the compensation event procedure.

In general, the ECC assumes that actions will be undertaken as required by the contract and that there are no fail-safes for not doing what is required. However, as the compensation event procedure is critical to the successful outcome of

Table 6.5 Decisions to be made by the *Project Manager* during the compensation event procedure

Clause	Decisions to be made by the *Project Manager*
61.1	▪ Did the event originate with the *Project Manager* or the *Supervisor* giving an instruction or notification, issuing a certificate, or changing an earlier decision?
61.3	▪ Has the *Project Manager* already notified the event to the *Contractor*? ▪ Has it been less than 8 weeks since the *Contractor* became aware of the event?
61.4	▪ Did the event arise from a fault of the *Contractor*? ▪ Has the event happened yet? ▪ Is the event expected to happen? ▪ Does the event have an effect upon Defined Cost, Completion or meeting a Key Date? ▪ Is the event one of the compensation events stated in the contract?
61.4	▪ Are the Prices, Completion Date and Key Dates going to be changed? ▪ Has the *Project Manager* instructed the *Contractor* to submit quotations? ▪ Did the *Project Manager* fail to reply to the *Contractor's* notification within the time allowed, even if the *Contractor* notified the *Project Manager* of this failure?
61.5	▪ Did the *Contractor* give an early warning that an experienced contractor could have given?
61.6	▪ Are the effects of a compensation event too uncertain to be forecast reasonably? ▪ Were any of these assumptions later found to have been wrong?
61.7	▪ Has the Defects Certificate been issued?
62.1	▪ Are alternative quotations required?

62.3	▪ Is the *Project Manager's* response to a quotation
	  – an instruction to submit a revised quotation or
	  – an acceptance of the quotation or
	  – a notification that they will be making their own assessment?

62.4	▪ Is a revised quotation needed?
62.5	▪ Is a longer time period needed for the *Contractor* to submit quotations or for the *Project Manager* to reply to a quotation?
62.2	▪ Did the *Project Manager* fail to reply to the *Contractor's* quotation within the time allowed, even if the *Contractor* notified the *Project Manager* of this failure?

63.1, 63.5	▪ Does the quotation include
	  – the effect of the compensation event on
	    ○ the actual Defined Cost of the work done by the dividing date
	    ○ the forecast Defined Cost of the work not yet done by the dividing date and
	    ○ the resulting Fee
	  – any delay to the planned Completion Date
	  – delay to the Key Date?

63.7	▪ Is the event to be assessed as if the *Contractor* had given an early warning?
63.8	▪ Does the assessment include time risk allowances for cost and time for matters that have a significant chance of occurring and are at the *Contractor's* risk under the contract?
63.10	▪ Is the compensation event an instruction to resolve an ambiguity or inconsistency?
63.11	▪ Is a correction to the Condition for a Key Date required?
64.1	▪ Has the *Contractor* submitted their quotation with details within the time allowed? ▪ Has the *Contractor* assessed the event correctly? ▪ At the time the *Contractor* submitted their quotation, did they also submit a programme or alterations to the programme? ▪ At the time the *Contractor* submitted their quotation, had the *Project Manager* accepted the *Contractor's* latest programme?

Clause	Decisions to be made by the *Project Manager*
64.2	■ Is there an Accepted Programme? ■ Has the *Contractor* submitted a programme or alterations to a programme for acceptance as required by the contract?
66.1	■ Has the *Project Manager* 　– notified their acceptance of the *Contractor's* quotation, or 　– notified the *Contractor* that they will do their own assessment? ■ Has a quotation been treated as having been accepted by the *Project Manager*?

the contract, there are three places where the procedure does include a fail-safe – these will ensure that the procedure moves on and reaches a conclusion even if the *Project Manager* does not do what they are supposed to do within the time allowed:

■ Deemed acceptance that the event is a compensation event and deemed instruction to submit quotations (clause 61.4): the *Project Manager* does not notify their decision within the time allowed.
■ Deemed acceptance by the *Project Manager* of a *Contractor's* quotation (clause 62.6): the *Project Manager* does not reply to a quotation within the time allowed.
■ Deemed acceptance by the *Project Manager* of a *Contractor's* quotation (clause 64.4): the *Project Manager* does not assess a compensation event within the time allowed.

Figure 6.2 outlines the procedures for deemed acceptance by the *Project Manager* in the compensation event procedure.

Figure 6.2 Deemed acceptance by the *Project Manager*

Clause 61.4

The *Project Manager* does not notify a decision within the time allowed

⬇

The *Contractor* notifies the *Project Manager* of their failure (to notify a decision within the time allowed)

⬇

The *Project Manager* does not reply within 2 weeks of this notification

⬇

The notification is treated as acceptance by the *Project Manager* that the event is a compensation event and an instruction to submit quotations

Clause 62.6

The *Project Manager* does not reply to a quotation within the time allowed

⬇

The *Contractor* notifies the *Project Manager* of their failure (to reply within the time allowed) and states which quotation they propose to be accepted

⬇

The *Project Manager* does not reply within 2 weeks of this notification

⬇

If the quotation is not for a proposed instruction or proposed changed decision, then the *Contractor's* notification is treated as acceptance of the quotation by the *Project Manager*

Clause 64.4

The *Project Manager* does not assess a compensation event within the time allowed

⬇

The *Contractor* notifies the *Project Manager* of their failure (to assess within the time allowed) and states which quotation they propose to be accepted

⬇

The *Project Manager* does not reply within 2 weeks of this notification

⬇

The notification is treated as acceptance by the *Project Manager* of the *Contractor's* quotation

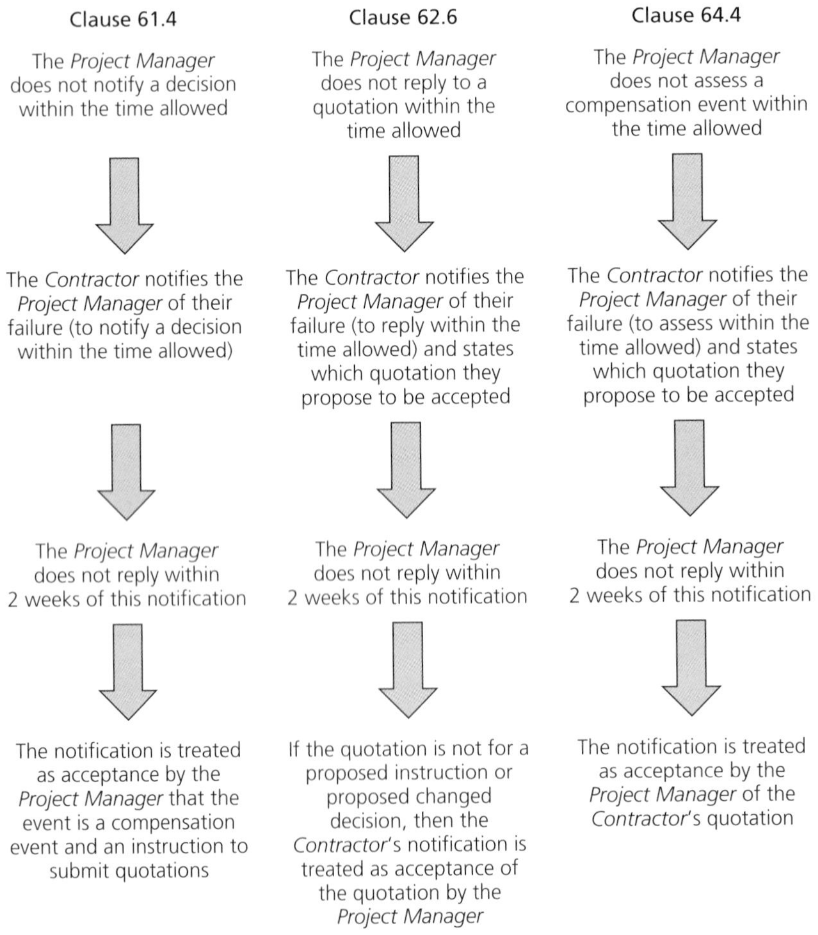

6.6.   First marking of Equipment, Plant and Materials

The *Project Manager* is not directly involved in the marking of Equipment, Plant and Materials that are outside the Working Areas. It is the *Supervisor* who marks them for payment, and if the contract identifies them for payment, the *Project Manager* includes them in the amount due.

Any requirements for marking of Equipment, Plant and Materials should be identified and discussed at regular intervals by the *Project Manager* and *Supervisor*. They should also agree how they will keep each other informed. Appendix 18 provides an agenda for monthly meetings between the *Project Manager* and the *Supervisor*.

6.7   First proposed instruction

If the *Client* or the *Project Manager* is considering making a change in the project, the *Project Manager* has the facility to instruct the *Contractor* to submit a quotation for a proposed instruction (clause 65.1). The instruction must also state the date by when the *Project Manager* thinks the proposed instruction may be given, and this adds a bit of reality – the *Project Manager* has to give some thought with regards to where the change fits into the Accepted Programme. The ensuing procedure is similar to the procedure for a compensation event:

- The *Contractor* submits quotations for the proposed instruction within 3 weeks of being instructed to do so.
- The quotation is assessed in the same way as a compensation event would be assessed.
- The *Project Manager* replies to the quotation (by the date when the proposed instruction may be given, and this date was included in the instruction to submit quotations),
  - requiring a revised quotation, with reasons, or
  - accepting the quotation and, at the same time, issuing the instruction and notifying a compensation event or
  - not accepting the quotation (in this case the *Project Manager* can still issue the instruction, notify a compensation event, and instruct the *Contractor* to submit a quotation for the compensation event; under clause 62.1, the *Project Manager* may elicit support from the *Contractor* by discussing different ways of dealing with the compensation event).

Section 7

NEC4: The Role of the *Project Manager*
ISBN 978-0-7277-6353-2

Chronological management procedures: second and later months

The meetings outlined at the start of the project will continue throughout the contract.

7.1. Second and later (revised) programmes submitted by the *Contractor*

The *Contractor* is required to submit programmes at regular intervals throughout the contract period. As well as the elements that must be shown on each programme under clause 31.2 and discussed in Section 6.2 of this book, the elements to be included in revised programmes are provided for in clause 32.1.

7.1.1 Information that the *Contractor* must include in their revised programmes

Once the *Project Manager* has accepted the first programme submitted to them by the *Contractor* for acceptance, that programme becomes the Accepted Programme. All subsequent programmes that are submitted to the *Project Manager* by the *Contractor* for acceptance, and that the *Project Manager* accepts, become the Accepted Programme, and they supersede all previous Accepted Programmes (clause 11.2(1)). Therefore, the second and subsequent programmes submitted will be based on the information provided in the first or previous programme, rather than the information that was submitted at the tender stage.

The *Project Manager* may also use other documentation and information when reviewing second and subsequent programmes:

- their own knowledge of progress on the Site
- the *Supervisor's* site diaries
- the *Supervisor's* Defects log, including notified Defects, Defects that have already been corrected and Defects that must be completed before Completion can be reached
- the *Contractor's* quotations for implemented compensation events, showing the forecast effect of the compensation event on the Completion Date, the Key Dates and the remaining work.

Sometimes the Scope includes a requirement for the *Contractor* to provide short-term or 'look ahead' programmes (e.g. a 4-week rolling programme) in addition to the Accepted Programme. These programmes enable a focus on the short-term activities required to maintain progress.

Appendix 19 provides a checklist of the dates/points in time and information that the *Contractor* must include in subsequent programmes submitted to the *Project Manager* for acceptance.

The regular submission of programmes to the *Project Manager* for acceptance is not affected by the submission of programmes at other times for specific reasons. The *Contractor* must submit revised programmes at the interval stated in the Contract Data, whether or not it has just submitted a programme to the *Project Manager* for acceptance as part of a compensation event, in response to an instruction to do so, or as a quotation for acceleration. However, the *Project Manager* can relax the requirement if they are so inclined. This is not specifically provided for by the ECC, but common sense dictates that the *Project Manager* will do no harm by suggesting to the *Contractor* that they need not submit a programme for acceptance if just 2 days ago the *Project Manager* accepted a quotation for a compensation event and the Accepted Programme was revised as a result.

7.1.2 The Accepted Programme: relationship with the Scope and other contract documents

The Scope and the Contract Data impact on the Accepted Programme; the Scope in particular should contain information that is required to be shown on the Accepted Programme. When the *Project Manager* reviews the *Contractor's* submitted programme, they should have the other parts of the contract to hand so that they can check that the required information has been shown on the programme.

Appendix 20 provides a checklist for comparing the requirements of the Scope and other contract documents with the *Contractor's* programme submitted for acceptance.

7.1.2.1 The programme on a wall

An Accepted Programme can contain thousands of activities, and therefore it can be difficult to understand what is happening at any one time. It is useful for the *Project Manager* to consider having a wallchart, as shown in Figure 7.1, which shows key programme information and can give an overview as to whether the time aspects of a project are going to be met. If it looks as though the planned date for completion of an activity will not be met, then the planner can roll down the programme into greater detail so that the underlying cause of delay and disruption can be identified. A risk reduction meeting can then be held to plan how to deal with the delay and disruption.

7.2. Assessment of the amount due: later months

Assessments of the amount due take place at the intervals stated in the Contract Data throughout the contract period. Cost-based contracts – those using main Option C, D or E – require careful monitoring of the actual Defined Cost. The *Project Manager* will need to check that the *Contractor* is controlling and managing Defined Cost in accordance with the contract.

The *Project Manager* of an Option C, D or E contract will want to set up procedures that facilitate an easier assessment, such as

- agreeing with the *Contractor* what details will be included in the applications for payment they submit every month, including those details already required by the Scope
- considering whether a part application every week or other period of time would be of benefit
- adopting communications that would facilitate a quick and realistic agreement of what work is coming up for which the *Contractor* will have paid out
- agreeing how information is to be provided by the *Contractor* so as to facilitate a quick review, for example
  - people costs should be presented with evidence of the costs described in items 11 to 13 of the SCC and, separately, costs described in item 14 of the SCC, to be followed in a similar manner with costs of Equipment supported by evidence as described in items 21 to 28 of the SCC
  - costs should be presented per activity, so that the activities undertaken for the *works* are separated out by their description and the people, Equipment, Plant and Materials costs and so on are all provided in bite-sized chunks that are easier to assimilate and can be mentally or physically compared with similar activities on other projects or earlier on the same project
- taking part in a joint visual inspection of the *works*, where the *Contractor* can point out aspects of the *works* that may be relevant to the application for payment, the *Supervisor* can refer to any Defects and how they will affect the *works* and payment, a planner can provide an overview of what work is to come next, and others involved in the contract also have the opportunity to contribute to the joint understanding of the project.

7.2.1 When do later assessments take place?

Assessments of the amount due that take place after the first assessment of the amount due are required to take place at the end of each *assessment interval* until the *Supervisor* issues the Defects Certificate or the *Project Manager* issues a termination certificate.

The *Project Manager* will understand the following:

- many of the assessments that take place after Completion and before the Defects Certificate/termination certificate may amount to zero
- assessments that take place after Completion may include assessments of costs that arise due to access issues.

7.2.2 What the *Project Manager* needs to know to make an assessment of the amount due

If a later assessment date is approaching, the *Project Manager* needs to have the following information:

- The date of the most recent assessment – the *Project Manager* should check the project wallchart, their diary or their payment certificates.
- The *assessment interval* – found in CD1 by the *Client* under section 5 ('Payment'). This is usually expressed as a number of weeks, but cannot be more than 5 weeks. As an alternative to providing an *assessment interval*, the

Figure 7.1 A wallchart showing progress

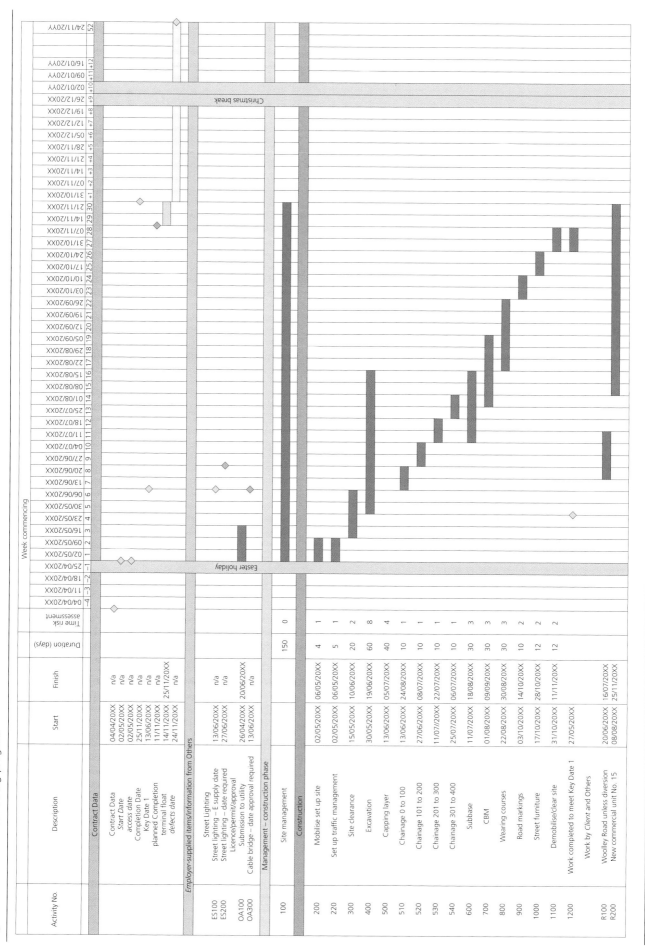

Contract Data may provide a schedule of assessment dates or it may describe an interval of time, such as the time period between the previous assessment date and the last Thursday of each calendar month.

- The next assessment date – it is likely that the *Project Manager* will have scheduled all the assessment dates over the period of the contract and marked them on their calendar or electronic device so that they know when the next date is due and can plan for it as required.

7.2.3 Undertaking later assessments

The *Project Manager's* assessment of the amount due is probably the most important activity that they will provide for the contract and the *Client*. They will undertake as much preparation as they can if they are to reduce the burden imposed by assessments for cost-based contracts. Appendix 21 provides a checklist of the information that the *Project Manager* needs to collect before undertaking an assessment of the amount due.

The effect of compensation events will need to be considered for all assessments undertaken prior to the Defects Certificate, but it will be most important in the assessments between the first assessment and an assessment that takes place at or just after Completion. At any point in time, the *Client* should be able to access the Price for Work Done to Date to understand the cost of the *works* up to that point. Unlike traditional contracts, where claims can be made at the end of the contract, the ECC requires a swift and effective response to changes so that the cost of the contract is up-to-date at all times, with the exception of notified compensation events that have not yet been implemented. This procedure is vital for the cost-based Option C, D or E contracts.

As the assessments of the amount due take place regularly, the *Project Manager* will develop their own rhythm for their assessments. They may choose to mark certain assessment activities in their electronic diary as a reminder of the *assessment interval*, and the *Project Manager* will take the following into consideration:

- application for payment by the *Contractor*
- quantity assessment and lump sum assessment
- Defects
- retention
- sectional Completion
- any Key Performance Indicators achieved.

The following are examples of the types of information, dates and so on that could be marked on a wallchart or in an electronic project diary (see also Figure 7.2):

- The date of the next assessment is Thursday 29 September 20XX (the last Thursday of the month).
- The *Contractor's* application for payment is therefore due the previous Monday: 26 September 20XX (as required by the Scope).
- The *Project Manager* will assess the quality of work that the *Contractor* has completed for each item in the Bill of Quantities and a proportion of each lump sum on Thursday 29 September 20XX. (This may mean that the quantity surveyor to whom the *Project Manager* has delegated their duties under clause 50.1 undertakes the assessment. Depending on the size of the project, the delegate *Project Manager* may undertake the assessment over a number of days, with the final result being due on Thursday 29 September 20XX.)
- On Friday 30 September 20XX the *Project Manager* will set up their spreadsheet for input, making sure that all aspects of the Price for Work Done to Date are included, such as
  - the quantity multiplied by the rate (without Defects)
  - the proportion of the lump sum (without Defects)
  - conversion to different currencies (Option X3)
  - the delay damages or bonus for early Completion for any completed *section* of the *works* (Option X5)
  - repayment of any advanced payment (Option X14)
  - retention (Option X16)
  - low-performance damages (Option X17)
  - measurement against a Key Performance Indicator (Option X20).
- On Monday 3 October 20XX the *Project Manager* will issue their payment certificate to the *Contractor* and the *Client* (the last day for issuing the payment certificate is Wednesday 5 October 20XX, giving the *Project Manager* 2 days' float).

7.2.3.1 Options A to F

The assessment takes place in the same way as for the first assessment.

Figure 7.2 Timeline of the payment procedure

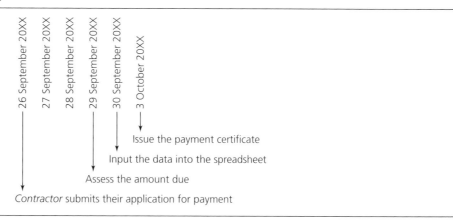

7.2.3.2 Assessments that take place outside regular assessments

Several clauses other than those in core clause section 5 require the *Project Manager* to make assessments of amounts due. Appendix 22 provides a checklist for assessments that take place outside core clause section 5.

7.3.   Tests and inspections

There is no contractual requirement for the *Supervisor* to keep the *Project Manager* in the loop about tests/inspections.

Tests and inspections may affect the project in a number of ways, as shown in Table 7.1.

7.4.   Quality management

Quality management is mainly the domain of the *Supervisor*, but the *Project Manager* is there to ensure that the *Client's* quality objectives for a project are met and that the *Contractor* adheres to their own quality plan.

Table 7.1 Effect of tests and inspections on clauses in the ECC

Issue	Clauses/conditions affected by tests and inspections
Defects	▪ A test or inspection may identify a Defect (clause 11.2(6)).
Completion	▪ The Scope may require certain tests and inspections to have been successfully completed before Completion can be certified. ▪ May prevent Completion being achieved (clause 11.2(2)).
Payment	▪ The *Project Manager* assesses any costs incurred by the *Client* in repeating tests or inspections after a Defect is found (clause 41.6). ▪ A payment is conditional upon a test or inspection being successful (clause 41.5). ▪ Correcting Defects (clause 44).
Compensation events	▪ *Client* does not provide something as shown on the Accepted Programme (clause 60.1(3)). ▪ The *Supervisor* instructs the *Contractor* to search and no Defect is found unless the search is needed only because the *Contractor* gave insufficient notice of doing work obstructing a required test or inspection (clause 60.1(10)). ▪ A test or inspection done by the *Supervisor* causes unnecessary delay (clause 60.1(11)). ▪ The *Client* does not provide materials, facilities and samples for tests and inspections as stated in the Scope (clause 60.1(16)).
Plant and Materials	▪ To be tested and inspected before delivery (clause 42).
Client dates to provide things	▪ Dates for *Client* to provide materials, facilities and samples for tests and inspections.
Keeping the *Client* informed	▪ Lack of knowledge about tests and inspections by the *Project Manager* will affect their ability to keep the *Client* informed.

Table 7.2 Aspects of quality management in the project

Issue	Quality management aspects
Scope	■ Is the *Contractor* complying with the Scope?
Quality management system	■ Is the *Contractor* operating a quality management system that complies with the requirements stated in the Scope?

Workmanship	■ Are there any concerns as to the general quality of workmanship
	  – by the *Contractor*?
	  – by their Subcontractors?
	■ Is the *Contractor's* quality plan being adhered to?
	■ Are Defects being notified?
	■ What is the level of Defects? (Is it excessive?)
	■ Are there any Defects that could delay or may prevent Completion being achieved (clause 11.2(2))?

Subcontracting/suppliers	■ Are there any concerns about the *Contractor's* Subcontractors or suppliers?
Plant and Materials	■ Are there any concerns as to the quality of Plant and Materials? ■ Have the Plant and Materials been tested or inspected before delivery as required by the Scope? ■ Are there excessive Plant and Materials on Site?
Equipment	■ Is Equipment being used safely? ■ Is Equipment being well maintained? ■ Are the levels of Equipment on Site adequate?

Quality management may affect the project in a number of ways, as shown in Table 7.2.

The *Project Manager* needs to know and be kept informed about all quality aspects of the project for the reasons identified in Table 7.2. The *Project Manager* should take note of any emerging patterns that may be obvious as to quality management aspects of the project. The *Project Manager* will need to set up their communication protocol with the *Supervisor* and hold regular meetings with them. Examples of agendas for meetings with the *Supervisor* are given in Appendices 5 and 11.

7.5. Dispute resolution

NEC contracts are designed to incentivise and promote good management practice and, by doing so, reduce the number of disputes that may arise on a project.

The key to avoiding disputes is for the *Project Manager* to follow clauses 10.1 and 10.2 to 'act as stated' and 'in a spirit of mutual trust and co-operation', respectively.

The *Project Manager* should therefore ensure that they

■ follow the management processes in the contract
■ give reasons for their actions
■ make decisions in a timely manner.

If issues or disputes do arise, then the *Project Manager* should strive to be proactive in discussing and understanding these, leaving no stone unturned. A *Project Manager* who finds themselves in a dispute must be able to stand in front of the *Senior Representatives*, an *Adjudicator* or a Dispute Avoidance Board and justify their actions and decisions.

The upshot of this for the *Project Manager* is that they should keep themselves well informed on all aspects of the project and should have good sources of information at hand so that they can make informed decisions.

7.5.1 Options W1, W2 and W3

The ECC includes three alternative procedures for dispute resolution, one of which is selected by the *Client* for each project:

- Option W1 is to be used where adjudication is the preferred route of dispute resolution, but the Housing Grants, Construction and Regeneration Act 1996 (HGCR Act) does not apply; that is, in countries other than the UK, and in the UK where the contract is not a construction contract as defined in the HGCR Act.
- Option W2 is to be used in the UK where the HGCR Act applies to the contract. Note that the Local Democracy, Economic Development and Construction Act 2009 amends some parts of the HGCR Act, and NEC4 has been amended accordingly.
- Option W3 is to be used where the Parties to a contract want to refer a disputed matter in the first instance to a Dispute Avoidance Board, whose task is to keep up to date with the progress of the project and to seek to avoid disputes before they happen.

Once a dispute has been referred to the first level of dispute resolution/avoidance, the Parties are required to follow the procedure described, and the *Project Manager* can make no further decisions. The *Project Manager* may be required to provide evidence for the dispute. The *Contractor* and the *Project Manager* will continue with the business of Providing the Works even if a dispute has been referred.

It should be noted that the NEC contracts are used internationally, and as such each country will have its own legal system in relation to the resolution of disputes.

Section 8

NEC4: The Role of the *Project Manager*
ISBN 978-0-7277-6353-2

ICE Publishing: All rights reserved
http://dx.doi.org/10.1680/nectrpm.63532.069

Completion and take over

The dates of Completion and take over are both decided by the *Project Manager* – see Table 8.1.

Table 8.1 The *Project Manager's* decisions on the dates of Completion and take over

Issue	Definitions and the *Project Manager's* decisions
Completion	Defined in clause 11.2(2). Completion occurs when the *Contractor* has done all the work required and the facility can be used. The *Project Manager* decides when Completion has occurred.
The Completion Date	Defined in clause 11.2(3). Refers directly to the *completion date* stated in the Contract Data as adjusted for compensation events.
The *completion date*	The *completion date* for the whole of the *works* can be found in either CD1 or CD2.
The date of Completion	The date of Completion is decided by the *Project Manager* and may be different from the Completion Date.
Take over	Take over occurs when the *Client* starts to use the *works*, which usually takes place after Completion.
Who is involved	Mostly the *Project Manager* but also the *Client* and the *Contractor*.
Involvement of the *Project Manager*	The *Project Manager* ■ certifies when Completion has taken place ■ decides the date of Completion ■ keeps track of the Completion Date on the Accepted Programme ■ certifies take over.

8.1. Reaching Completion

The definition for Completion (clause 11.2(2)) directly refers to the Scope. When the *Project Manager* is deciding whether Completion has taken place, they will use the Scope and the list of notified Defects provided by the *Supervisor* (note that the ECC does not require the *Supervisor* to provide the list of Defects, and therefore the *Client* may want to ensure that the *Supervisor's* personal contract provides for this).

■ Clause 11.2(2)   The Scope states what work must be done by the Completion Date. Completion can only be reached when there are no Defects that would prevent the *Client* from using the *works* and Others from doing their work.

■ Clause 30.2   The *Project Manager* decides the date of Completion.

■ Clause 30.2   The *Project Manager* certifies Completion within 1 week of Completion.

8.1.1   The Scope

The Scope should provide a description of what work has to be done by Completion. The *Project Manager* will refer to this description when deciding whether Completion has taken place.

There is no 'practical completion' in the ECC (or 'mechanical completion' or 'substantial completion'), and the decision about whether Completion has taken place should be an objective one. The decision will be led by the *Contractor* having completed all of the work that the Scope states they are to do by the Completion Date and having corrected any notified Defects that would have prevented the *Client* from using the *works* and Others doing their work.

It is possible that the description of the work to be done by the Completion Date will not be found in one place in the Scope.

8.1.2   Defects that would prevent the *Client* using the *works*

The *Project Manager* may need to refer to the *Supervisor* for this aspect of Completion. The *Supervisor* will be able to provide the *Project Manager* with a list of uncorrected Defects as Completion nears. Since Completion is decided by the *Project Manager*, it is the *Project Manager* who decides whether any of the uncorrected Defects will prevent the *Client* from using the *works*. The *Project Manager* must therefore be privy to the *Client's* intended use of the *works*.

8.1.3   Defects that would prevent Others from doing their work

The Scope describes the *works*, and should include statements about what work will be done within, on or to the completed *works* by Others (e.g. other contractors). However, if the *Project Manager* is required to make a decision about whether a Defect could prevent work by Others, then they must have knowledge about what work will be carried out by Others after Completion. The *Client* will need to make sure that this communication takes place – whether in the *Project Manager's* personal contract or in some other communication provided to the *Project Manager* – in sufficient time for the *Project Manager* to be able to make a decision on whether Completion has taken place.

> The *Project Manager* decides the date of Completion. The date of Completion might not be the same as the Completion Date.

8.1.4   The certificate of Completion

Once the *Project Manager* has decided the date of Completion, they certify Completion within 1 week of that date. The Completion certificate is issued to the *Client* and the *Contractor* (clause 13.6). An example of a Completion certificate can be found in the ECC Guidance Notes.

8.2.   Consequences of Completion

All of the comments in this section referring to Completion apply also to any *section* of the *works*, as described using secondary Option X5 (sectional Completion), unless otherwise stated.

The majority of the milestone points in the ECC refer to the *defects date* or take over, rather than to Completion. The most obvious and immediate consequence of Completion is that take over occurs after Completion. (Section 8.4 below provides a discussion of take over.)

Other consequences are as follows:

- **Option X1, 'Price adjustment for inflation'.** The Price Adjustment Factor is calculated at the Completion Date for the whole of the *works* and then used to calculate any price adjustments after this date (clause X1.2). The calculation takes place specifically at the Completion Date for the whole of the *works*; therefore, this clause does not apply to any Completion Dates for *sections* of the *works*. It also refers to the Completion Date rather than the 'date of Completion', so the calculation takes place when Completion is programmed to take place, rather than when it actually takes place (on the basis that the two dates might not coincide).
- **Option X5, 'Sectional Completion'.** If Option X5 is part of the contract, then references to Completion and the Completion Date apply to any *section* of the *works* as well as to the whole of the *works*. Therefore,
  - the *Project Manager* decides the date of Completion for the *sections* of the *works* using the same method described above
  - if Completion of a *section* takes place before the Completion Date for that *section*, then the *Client* is not required to take over that *section* if it was stated in CD1 that they are not willing to do so, otherwise the *Client* takes over that *section* not later than 2 weeks after Completion.
- **Option X6, 'Bonus for early Completion'.** If Option X6 is used (but not if Option X5 is also used), the Contract Data entry is as follows:

> The bonus for the whole of the *works* is . . . . per day.

The *Contractor* is paid a bonus calculated at the rate stated in the Contract Data, for each day from the earlier of
– Completion of the *works* (the whole of the *works*) or
– take over of the *works* (the whole of the *works*)
until the Completion Date.

If Completion of the *works* is earlier than take over:	The *Contractor's* bonus is calculated from the date of Completion of the *works* (the date is certified by the *Project Manager* on the Completion certificate) to the Completion Date (stated on the Accepted Programme)
If take over of the *works* is earlier than Completion of the *works*:	The *Contractor's* bonus is calculated from the date of take over (the date is certified by the *Project Manager* on the take over certificate) to the Completion Date (stated on the Accepted Programme)

- **Option X7, 'Delay damages'.** Two possible scenarios are described here:
  - if Completion takes place after the Completion Date the *Contractor* pays delay damages from the Completion Date (stated on the Accepted Programme) until Completion (the date is certified by the *Project Manager* on the Completion certificate)
  - if take over takes place before Completion the *Contractor* pays delay damages from the Completion Date (stated on the Accepted Programme) until the date on which the *Client* takes over the *works* (the date is certified by the *Project Manager* on the take over certificate).
  If Options X5 and X7 are used together, then these actions also apply to *sections* of the *works*.
- **Option X16, 'Retention'.** Two possible scenarios are described here:
  - if Completion of the whole of the *works* takes place before take over of the whole of the *works*, then Completion marks the end of the period during which the *Client* retains the *retention percentage* in each Price for Work Done to Date, and the amount of retention is halved in the assessment made at Completion of the whole of the *works*
  - if the date on which the *Client* takes over the whole of the *works* occurs before Completion of the whole of the *works*, the take over date marks the end of the period during which the *Client* retains the *retention percentage* in each Price for Work Done to Date, and the amount of retention is halved in the next assessment after the *Client* has taken over the whole of the *works*.
  Neither of these two scenarios applies to *sections* of the *works*.
- **Option X22, 'Early *Contractor* involvement (used only with Options C and E)'.** The *Project Manager* makes a preliminary assessment of the *budget incentive* at Completion of the whole of the *works* and includes this in the amount due following Completion of the whole of the *works*.

8.3. Assessment at Completion

The ECC does not require a special assessment of the amount due to take place at Completion – any assessment is undertaken at the end of the *assessment interval* during which Completion took place. Unlike in traditional contracts, the assessment that takes place at (or just after) Completion should not take much longer than any of the previous assessments. If the contract has been managed as required by the ECC, there is no 'final account', and the assessment that takes place at Completion is complicated only by some of the secondary Options. Appendix 23 provides a checklist of what needs to be collected to undertake the assessment of the amount due at Completion of the whole of the *works* (for Options A to F).

In assessing the amount due at Completion of the whole of the *works* the *Project Manager* needs to consider a number of other things, such as any records (operation and maintenance manuals, as-built drawings, etc.) that are required to be provided before Completion is reached. They should also review the list of notified Defects, to understand which Defects still have to be corrected and whether these will be corrected after Completion. The cost-based Option C, D and F contracts disallow the cost of correcting Defects after Completion.

Appendix 24 provides a checklist for the assessment of the amount due at Completion of the whole of the *works*.

8.4. Take over

Take over of the *works* can take place before the Completion Date if the *Client* is willing, but it is more likely to take place after Completion. Unlike Completion, the status of take over is not decided by the *Project Manager*; rather, it is linked to Completion through time, as it takes place no later than 2 weeks after Completion. Take over signifies the transfer of the project back to the *Client*.

The most common scenario is that Completion will be reached and certified and then take over will take place. In this case, the actions are as follows:

1 completion is reached on or before the Completion Date
2 the *Project Manager* certifies Completion **within 1 week** of the date of Completion
3 take over takes place **not more than 2 weeks** after Completion
4 the *Project Manager* certifies take over **within 1 week** of the date of take over.

8.4.1 Scenarios linking Completion, the Completion Date and take over

The *Client* may choose whether to take over the *works* before the Completion Date, but the choice must be made at the start of the contract through the use of the relevant optional statement in CD1.

CD1 statement if the *Client* is not willing to take over the *works* before the Completion Date

▪ The *Client* is not willing to take over the *works* before the Completion Date.

The *Client* still has the option to use the *works* before Completion (which may be on or before the Completion Date). If the *Client* uses the *works* before Completion, take over takes place immediately, unless

▪ the reason for using the *works* before Completion is stated in the Scope or
▪ the use suits the *Contractor's* method of working.

If the *Client* has not stated in the Contract Data that it is unwilling to take over the *works* before the Completion Date, the implication is that if the *works* were to reach Completion before the Completion Date, the *Client* would be willing to take over the *works* 2 weeks after Completion, even if that date were to be before the Completion Date.

The following describes the various scenarios that may be experienced. The only scenario that would change if the Contract Data has not stated that the *Client* is not willing to take over the *works* before the Completion Date is scenario 2.

▪ **Scenario 1**
 – The *Client* is not willing to take over the *works* before the Completion Date.
 – Completion takes place 1 week before the Completion Date.

1 July 20XX	Completion
7 July 20XX	Completion Date
15 July 20XX	Take over (takes place no later than 2 weeks after Completion)
22 July 20XX	The *Project Manager* issues the take over certificate (within 1 week of the date of take over)

▪ **Scenario 2**
 – The *Client* is not willing to take over the *works* before the Completion Date.
 – Completion takes place 3 weeks before the Completion Date.

1 July 20XX	Completion
15 July 20XX	Take over would generally take place no later than 2 weeks after Completion, but, since the *Client* is not willing to take over the *works* before the Completion Date, take over does not happen at this time
22 July 20XX	Completion Date
5 August 20XX	Take over (takes place no later than 2 weeks after Completion)
12 August 20XX	The *Project Manager* issues the take over certificate (within 1 week of the date of take over)

In this scenario, if the *Client* had not indicated that they were not willing to take over the *works* before the Completion Date, the dates would be as follows:

1 July 20XX	Completion
15 July 20XX	Take over (takes place no later than 2 weeks after Completion)
22 July 20XX	Completion Date
22 July 20XX	The *Project Manager* issues the take over certificate (within 1 week of the date of take over)

■ **Scenario 3**
- The *Client* is not willing to take over the *works* before the Completion Date.
- Completion takes place on the Completion Date.

15 July 20XX	Completion
15 July 20XX	Completion Date
29 July 20XX	Take over (takes place no later than 2 weeks after Completion)
5 August 20XX	The *Project Manager* issues the take over certificate (within 1 week of the date of take over)

■ **Scenario 4**
- The *Client* is not willing to take over the *works* before the Completion Date.
- Completion takes place 1 week after the Completion Date.

1 July 20XX	Completion Date
8 July 20XX	Completion
22 July 20XX	Take over (takes place no later than 2 weeks after Completion)
29 July 20XX	The *Project Manager* issues the take over certificate (within 1 week of the date of take over)

■ **Scenario 5**
- The *Client* is not willing to take over the *works* before the Completion Date.
- The *Client* starts to use the *works*, and the reason for using the *works* is stated in the Scope.
- Completion takes place 1 week before the Completion Date.

29 June 20XX	The *Client* uses the *works*
1 July 20XX	Completion
7 July 20XX	Completion Date
15 July 20XX	Take over (takes place no later than 2 weeks after Completion because the *Client's* use of the *works* was for a reason stated in the Scope)
22 July 20XX	The *Project Manager* issues the take over certificate (within 1 week of the date of take over)

Completion is because it suits the way the *Contractor* is working.

■ **Scenario 6**
- The *Client* is not willing to take over the *works* before the Completion Date.
- The *Client* starts to use part of the *works* before the Completion Date, and the reason given for this use is not included in the Scope, nor does it suit the *Contractor's* way of working.
- Completion takes place 1 week before the Completion Date.

29 June 20XX	The *Client* uses part of the *works*
29 June 20XX	Take over (on use) even though CD1 provided that the *Client* was not willing to take over the *works* before the Completion Date
1 July 20XX	Completion
6 July 20XX	The *Project Manager* certifies take over within 1 week
6 July 20XX	The *Project Manager* notifies the *Contractor* of a compensation event under clause 60.1(15) and instructs the *Contractor* to submit quotations
8 July 20XX	Completion Date

In this scenario, the *Client's* use of a part of the *works* triggers an immediate take over, as the use is in contradiction to the statement included in the Contract Data. Since the use of the *works* is not provided for in the Scope or to suit the *Contractor's* way of working, take over takes place on use of the *works*. In addition, take over of a part of the *works* before both Completion and the Completion Date is a compensation event under clause 60.1(15), and so the *Project Manager* must notify a compensation event at the same time as they certify take over (however, if the *Client* takes over a part of the *works* and the *Contractor* is in delay, then this is not a compensation event).

8.4.2 The take over certificate
Once the *Project Manager* has decided the date of take over, they must certify take over within 1 week of that date. Their take over certificate is issued to the *Client* and the *Contractor* (clause 13.6). An example take over certificate is included in the ECC Guidance Notes.

8.4.3 Consequences of take over
The following actions must result from take over:

- If the *Contractor* requires access to the *works* to correct a Defect, then the *Project Manager* arranges access with the *Client*.
- Take over before Completion and the Completion Date is a compensation event under clause 60.1(15) (but note that the clause refers to the certification of take over rather than to take over itself).
- Loss of or wear or damage to the parts of the *works* taken over by the *Client* (with exceptions) is a *Client's* liability.
- Option X6, 'Bonus for early Completion' – if Completion takes place before take over, then the *Contractor's* bonus is calculated from the date of Completion (the date is certified by the *Project Manager* on the Completion certificate) to the Completion Date (stated on the Accepted Programme). If take over occurs before Completion, then the *Contractor's* bonus is calculated from the date of take over (the date is certified by the *Project Manager* on the take over certificate) to the Completion Date (stated on the Accepted Programme). If Options X5 and X6 are both part of the contract, these actions also apply to sections of the *works*.
- Option X7, 'Delay damages' – if Completion is after the Completion Date, the *Contractor* pays delay damages from the Completion Date to Completion (date certified by the *Project Manager* on the Completion Certificate).
- Option X16, 'Retention' – the retention amount is halved in the next assessment made after Completion of the whole of the *works* or in the next assessment after the *Client* has taken over the whole of the *works*.

Six checklists for Completion and take over are included in Appendix 25.

Section 9

NEC4: The Role of the *Project Manager*
ISBN 978-0-7277-6353-2

ICE Publishing: All rights reserved
http://dx.doi.org/10.1680/nectrpm.63532.077

After Completion and take over

The period of time between the date of Completion and the *defects date* primarily involves the *Supervisor*; however, the *Project Manager* still has a number of responsibilities.

- **What happens after Completion?** Apart from the immediate implications of reaching take over, the activities that take place after Completion include the following:
  - The correction of Defects that were notified before Completion.
  - The *Project Manager* arranges access with the *Client* for the *Contractor* to use part of the *works* to enable the *Contractor* to correct a Defect.
  - The *Project Manager* assesses the amount due at the next assessment date after Completion of the whole of the *works* and at the end of each *assessment interval* until the *Supervisor* issues the Defects Certificate.
  - Title to any remaining Plant and Materials passes back to the *Contractor*, as it is removed from the Working Areas when the *Contractor* leaves the Site or scales back their operations to concentrate on the correction of Defects.
  - Liabilities are still held by the *Client* and the *Contractor*.
- **Who is involved?** Primarily the *Supervisor*, but the *Project Manager* has some responsibilities, and needs to liaise with the *Supervisor* in regard to the progress on the correction of notified Defects. Any uncorrected Defects are assessed and taken into account in the assessment of the amount due at each assessment interval.

9.1. Defects

The *Contractor* is not obliged to correct notified Defects before Completion unless they are Defects that would prevent the *Client* from using the *works* or Others from doing their work and would therefore affect Completion (clause 11.2(2)). Instead, the *Contractor* is obliged to correct notified Defects before the end of their *defect correction period*. For Defects notified before Completion the *defect correction period* begins at Completion.

The *Project Manager* is only part of this process in the following instances:

- If, after take over, the *Contractor* needs access to and use of a part of the *works* in order to correct a Defect, it is the *Project Manager* who arranges for the *Client* to allow the *Contractor* access and use of the relevant part of the *works*. Then,
  - if the *Contractor* is not given access to the *works* before the *defects date* for this purpose, the *Project Manager* assesses the cost to the *Contractor* of correcting the Defect (clause 46.2) or
  - if, after the *Contractor* has been given access, it does not correct the Defect within its *defect correction period*, the *Project Manager* assesses the cost to the *Client* of having the Defect corrected by other people (clause 46.1).
- The *Project Manager* or *Contractor* may propose to the other that the Scope should be changed so that a Defect does not have to be corrected (clause 45.1). If the *Contractor* and *Project Manager* are prepared to consider the change, the *Contractor* submits a quotation for reduced Prices or an earlier Completion Date or both to the *Project Manager* for acceptance. If the *Project Manager* accepts the quotation, they give an instruction to change the Scope, the Prices, Key Dates and the Completion Date.
- The *Project Manager* should note that the quotation from the *Contractor* does not have to follow the requirements of core clause section 6 compensation events. The quotation provided by the *Contractor* is a commercial offer from the *Contractor* for not correcting a Defect.
- In considering the *Contractor's* quotation the *Project Manager* is likely to have discussed the proposed acceptance of a Defect with the *Client's* advisors and other members of the project team, such as the *Client's* designers, engineers and so on.

- The acceptance of a Defect may have capital and operational cost implications: for example, the equipment may be more expensive to operate or the life of the asset may be reduced. Some *Clients* have internal processes to follow in regard to derogation of standards or concession applications, especially in operational businesses such as railways or airports where health and safety are paramount.
- The acceptance of a Defect is likely to require checking and review by the *Client's* advisors and other consultants, which will incur additional costs. Some *Clients* include a requirement that the assessment of any acceptance of a Defect includes as a reduction to the total of the Prices of any additional *Client's* costs incurred in considering the proposal.

9.2. Use of the *works*

At take over the *Client* starts to use the *works*. The *Client* then becomes responsible for providing access so that the *Contractor* can correct notified Defects during the period from take over up until the end of the last *defect correction period*, which could run from the *defects date*.

The *Project Manager* arranges for the *Client* to allow access to and use of a part of the *works* that they have taken over if this is necessary for correcting a Defect. The *defect correction period* begins when the necessary access and use have been provided.

Section 10

NEC4: The Role of the *Project Manager*
ISBN 978-0-7277-6353-2

ICE Publishing: All rights reserved
http://dx.doi.org/10.1680/nectrpm.63532.081

After the *defects date*

The *defects date* marks the end of the project. The actual end date is formalised by the *Supervisor* in the Defects Certificate (Table 10.1).

Table 10.1 Key points about the Defects Certificate

What is it?	It marks the end of the project.
Who issues it?	The *Supervisor*.
What are the implications for the *Project Manager*?	The *Project Manager's* final assessment of the amount due takes place no later than 4 weeks after the issue of the Defects Certificate. The 4-week timescale provides time for the *Project Manager* to assess the cost of correcting Defects listed on the Defects Certificate and Option X17, 'Low performance damages'.

10.1. The *defects date*

The *defects date* is identified in CD1 in section 4, 'Quality management'. It is defined as a period of time after Completion of the whole of the *works*, for example

The period between Completion of the whole of the *works* and the *defects date* is 52 weeks

Table 10.2 outlines the clauses related to the *defects date*.

Table 10.2 Clauses related to the *defects date*

Clause	Provision
11.2(7)	Defects can only be notified up to the *defects date* and not beyond.
41.5	The *defects date* marks the point[a] when payment conditional on tests/inspections by the *Supervisor* and not carried out becomes due. [a]The later of the *defects date* and the end of the last *defect correction period*.
43.1	The *Supervisor* may instruct searches up to the *defects date*.
43.2	The *Supervisor* and *Contractor* notify Defects up to the *defects date*.
61.7	Compensation events are not notified after the issue of the Defects Certificate.

10.2. The Defects Certificate

The Defects Certificate is defined in clause 11.2(7). Table 10.3 sets out the linked clauses that tell us more about the Defects Certificate. The *Supervisor* issues the Defects Certificate at the *defects date* if there are no notified Defects, or otherwise at the earlier of the end of the last *defect correction period* and the date when all notified Defects have been corrected. Appendix 26 provides a checklist for the *defects date* and the issue of the Defects Certificate.

Table 10.3 Clauses about the Defects Certificate

Clause	Provision
11.2(7)	The Defects Certificate is either a list of Defects that the *Supervisor* has notified before the *defects date* which the *Contractor* has not corrected or, if there are no such Defects, a statement that there are none.
44.3	The *Supervisor* issues the Defects Certificate at the *defects date* if there are no notified Defects, or otherwise at the earlier of ■ the end of the last *defect correction period* ■ the date when all notified Defects have been corrected. The *Client's* rights in respect of a Defect which the *Supervisor* has not found or notified are not affected by the issue of the Defects Certificate.
50.1	An assessment of the amount due takes place at the end of each *assessment interval* until the *Supervisor* issues the Defects Certificate.
53.1	The final assessment of the amount due takes place no later than 4 weeks after the *Supervisor* issues the Defects Certificate.
80.1	The *Client's* liabilities are affected by the issue of the Defects Certificate: loss of or damage to the parts of the *works* taken over by the *Client*, except loss or damage occurring before the issue of the Defects Certificate which is due to ■ a Defect that existed at take over or ■ an event occurring before take over which was not itself a *Client's* liability or ■ the activities of the *Contractor* on Site after take over.
83.3	Insurance by the *Contractor* does not need to continue beyond the issue date of the Defects Certificate or a termination certificate being issued.
X16.2	Any remaining retention held is released in the assessment of the amount due after the Defects Certificate is due to be issued.
X17.1	Any Defect included in the Defects Certificate may require payment by the *Contractor* if the Defect shows low performance and Option X17 applies to the contract.
X18.4	Option X18 affects Defects due to the *Contractor's* design and not listed in the Defects Certificate.
X20.1	The issue of the Defects Certificate marks the end of the *Contractor's* reporting obligations under Option X20.

10.3. Assessment of the final amount due

The assessment of the final amount due is the last assessment that the *Project Manager* will undertake for the project. It takes place no later than 4 weeks after the *Supervisor* has issued the Defects Certificate or 13 weeks after the *Project Manager* issues a termination certificate. For the purposes of this book, the final assessment is assumed to take place no later than 4 weeks after the *Supervisor* has issued the Defects Certificate.

Even if the *Project Manager* is no longer on Site regularly, they will receive a copy of the *Supervisor's* Defects Certificate (in accordance with clause 13.6), and this will prompt the final assessment. The primary need for this assessment is the release of the second half of the retention under secondary Options X16 (not for Option F). Any assessments under clauses not in core clause section 5 (such as clause 46, 'Uncorrected Defects') can take place at any point before the final assessment as well, as the *Project Manager's* regular (e.g. monthly) assessments take place right through the period from Completion until the *Supervisor* issues the Defects Certificate.

Regular assessments end when the *Supervisor* issues the Defects Certificate. The date of the final assessment, which must take place no later than 4 weeks after the *Supervisor* issues the Defects Certificate, can be decided by the *Project Manager*.

Appendix 27 describes the information that the *Project Manager* needs to consider for the final assessment, and Appendix 28 provides a checklist for the *Project Manager* to use when they undertake the final assessment.

Section 11

NEC4: The Role of the *Project Manager*
ISBN 978-0-7277-6353-2

ICE Publishing: All rights reserved
http://dx.doi.org/10.1680/nectrpm.63532.085

Post-project evaluation and learning from experience

Undertaking a post-project evaluation and putting in place the procedures to learn from experience are part of good project management.

11.1. Post-project evaluation

In the past, we completed projects and then undertook post-project reviews to look at what went well and what did not go well. By this time the horse had usually bolted. The NEC is all about active risk-based management: it has strong management processes that are all work-flowed on risk, change and programme management, which the contract requires to be implemented and used in real time. NEC contracts also lend themselves to the use of 'in the cloud' management systems.

As a result, the *Project Manager* is in a far better position to make informed choices throughout the project, as the NEC provides them with opportunities to make decisions that reduce, mitigate or avoid potential risks. If implemented properly, the NEC management processes enable real-time evaluation of issues and will make any post-project evaluation far more effective and efficient.

The *Project Manager* may be asked to run a post-project evaluation workshop. Table 11.1 sets out the key steps and issues for running such a workshop. To get the most out of these workshops the *Project Manager* should follow these steps, but they will also need to consider the following:

- **invite the right people** – people who have been involved throughout the project, and other key stakeholders
- **set the ground rules** – a workshop is not about casting blame, but is a chance to consider the experiences and learning points of those involved in the project, both as individuals and as organisations
- **encourage people to evaluate the project** – they should identify what went well, and what can be improved.

Table 11.1 Key steps in post-project evaluation

Step	Description	Comments

1	Information required for the post-project evaluation meeting	■ Performance against project objectives:
		  – time, cost, quality, etc.
		■ Issues that arose during the project:
		  – the Early Warning Register
		  – early warning notifications
		  – other issues
		  – changes in the project environment
		■ Why change occurred:
		  – objectives, scope, etc.
		■ The *Client's* satisfaction with the project
		■ Management's satisfaction with the project
		■ Effectiveness of the project management processes:
		  – communication
		  – early warnings
		  – the Early Warning Register
		  – the Accepted Programme
		■ Lessons learned

Step	Description	Comments
2	Meeting agenda	■ Purpose of the meeting ■ Outcomes ■ Project performance: – time, cost, quality and Key Dates ■ Project management ■ Issues and risks ■ Learning from experience
3	Meeting	Review and explore the project outcomes and learning through questions and facts ■ Were the project objectives met? ■ Did the project meet the cost/time and quality objectives? ■ What issues arose during the project? ■ How well did the management processes in the NEC operate? ■ Were the organisation's own internal processes used correctly?
4	Record outcomes	Record the findings and outcomes
5	Improvement plans	Implement improvement plans based on the findings of the post-project evaluation

11.2. Learning from experience

Whoever takes on the role of *Project Manager* should take the opportunity to critically analyse their own performance and to learn from their experiences on an ECC contract.

Learning from experience is an important part of BS ISO 44001, 'Collaborative business relationship management systems'.

NEC4: The Role of the *Project Manager*
ISBN 978-0-7277-6353-2

Summary

The role of the *Project Manager* on NEC4 ECC contracts carries with it great responsibility. It requires a broad range of competencies, skills, knowledge and experience. It requires interpersonal skills, the ability to communicate effectively, to listen, discuss, consult, show leadership and collaborate with all those involved on a project. In short, the *Project Manager* leads and sets the tone for a project.

NEC contracts are based on active risk-based management. Failure to follow and implement the management processes or to act as stated or in a spirit of mutual trust and co-operation will impact upon the successful outcome of a project.

NEC4 ECC has potential benefits but these can only be realised by implementing and using the management processes in the contract and the person who must lead and ensure this happens is the *Project Manager*.

Appendices

NEC4: The Role of the *Project Manager*
ISBN 978-0-7277-6353-2

ICE Publishing: All rights reserved
http://dx.doi.org/10.1680/nectrpm.63532.091

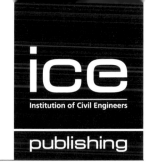

Appendix 1
Contract communications

Appendix 1A: Rules of communication
Communication must be as follows.

Clause	Description
13.1	In the *language of the contract*
13.2	Communication has effect when it is communicated through the communication system specified in the Scope Or If not through a communication system it takes effect when received at the last address notified by the recipient
13.3	Replied to within the *period for reply*
13.4	Replied to if it is a communication submitted or resubmitted by the *Contractor*
13.4	Accompanied by a reason if the reply is non-acceptance
13.5	The *Project Manager* may extend the period for reply
13.6	The *Project Manager* must issue their certificates to the *Client* and the *Contractor*
13.7	A notification or certificate required by this contract should be communicated separately from other communications
13.8	The *Project Manager* may withhold acceptance of a submission by the *Contractor*

Appendix 1B: List of communications required from the *Project Manager*

Clause	To whom	About what	Timing[a]
13.4	*Contractor*	A reply to a communication submitted or resubmitted to the *Project Manager* by the *Contractor*	Within the *period for reply*
14.1	*Contractor*	Acceptance of a communication	Within the period stated in the contract (or the *period for reply*)
14.2	*Contractor*	Delegate and cancel delegation of actions	
14.3	*Contractor*	Instruction changing the Scope or a Key Date	
15.1	*Contractor*	Early warning notification	
15.2	*Contractor*	Issues the first Early Warning Register	Within 1 week of the *starting date*
	Contractor	Instruction to attend an early warning meeting	
	Other people	Instruction to attend an early warning meeting	

Clause	To whom	About what	Timing[a]
15.4	*Contractor*	Issuing the revised Early Warning Register	Within 1 week after each early warning meeting
	Contractor	Instructs a change in the Scope at the same time as issuing the Early Warning Register	Within 1 week after each early warning meeting
16.2	*Contractor*	Reply to the *Contractor's* proposal to change the Scope provided by the *Client* in order to reduce the total of the Prices; and an instruction changing the Scope or an instruction to submit a quotation	Within 4 weeks of receiving the proposal
16.3	*Contractor*	Reply to the *Contractor's* proposal to add to the Working Areas – acceptance or non-acceptance with a reason	Within the *period for reply*
17.1	*Contractor*	Notification of ambiguity/inconsistency in or between contract documents	
	Contractor	Reply to the *Contractor's* notification of ambiguity/ inconsistency in or between contract documents	Within the *period for reply*
17.2	*Contractor*	Notification that the Scope includes an illegal or impossible requirement	
	Contractor	Instruction to change the Scope appropriately	
	Contractor	Reply to the *Contractor's* notification that the Scope includes an illegal or impossible requirement	Within the *period for reply*
	Contractor	Instruction to change the Scope appropriately	
19.1	*Contractor*	Instruction stating how the *Contractor* is to deal with the event	
21.2	*Contractor*	Reply to the *Contractor's* design particulars – acceptance or non-acceptance with a reason	Within the *period for reply*
23.1	*Contractor*	Instruction to submit the particulars of the design of an item of Equipment for acceptance	
	Contractor	Reply to the *Contractor's* design particulars for an item of Equipment – acceptance or non-acceptance with a reason	Within the *period for reply*
24.1	*Contractor*	Reply to the *Contractor's* submission of a proposed replacement person – acceptance or non-acceptance with a reason	Within the *period for reply*
24.2	*Contractor*	Instruction to remove an employee	
25.3	*Contractor*	Notification that the work does not meet the Condition stated for a Key Date by the date stated and that, as a result, the *Client* incurs additional cost[b]	
26.2	*Contractor*	Reply to the *Contractor's* submission of a proposed Subcontractor – acceptance or non-acceptance with a reason	Within the *period for reply*

Clause	To whom	About what	Timing[a]
26.3	*Contractor*	Reply to *Contractor's* submission of proposed subcontract documents, excepting pricing information, for each subcontract – acceptance or non-acceptance with a reason	Within the *period for reply*
	Contractor	Notification that no submission of proposed subcontract documents, excepting pricing information, is required[b]	
27.2	*Contractor*	Notification to provide access for Others to work being done and Plant and Materials being stored	
30.2	*Contractor* *Client*	Certificate of Completion	Within 1 week after Completion
31.3, 32.2	*Contractor*	Reply to *Contractor's* submission of a programme for acceptance – acceptance or non-acceptance with a reason	Within 2 weeks of submission
34.1	*Contractor*	Instruction to stop or not to start any work	
	Contractor	Instruction to start or restart any work	
35.3	*Contractor* *Client*	Certificate of take over	Within a week of the date of take over
36.1	*Contractor*	Proposal for an acceleration to achieve Completion before the Completion Date	
	Contractor	Reply that the *Contractor's* proposal for an acceleration to achieve Completion before the Completion Date will not be considered[b]	Within the *period for reply*
Or	*Contractor*	Instruction to submit a quotation for an acceleration, stating changes to Key Dates to be included in the quotation	
	Contractor	Reply to the *Contractor's* quotation – acceptance or non-acceptance with a reason	Within 3 weeks of submission of the quotation
36.3	*Contractor* *Client*	If a quotation for an acceleration is accepted, notification changing the Prices, the Completion Date and the Key Dates; and accepting the revised programme[b]	
40.2	*Contractor*	Reply to the *Contractor's* submission of a quality policy statement and a quality plan submitted for acceptance – acceptance or non-acceptance with a reason	Within the *period for reply*
	Contractor	Reply to the *Contractor's* submission of a changed quality plan submitted for acceptance – acceptance or non-acceptance with a reason	Within the *period for reply*
40.3	*Contractor*	Instruction to correct a failure to comply with the quality plan	
44.4	*Contractor*	Notification that the *Client* is allowing access to and use of a part of the *works* that has been taken over for the purpose of correcting a Defect[b]	

Clause	To whom	About what	Timing[a]
45.1	*Contractor*	Proposal to change the Scope so that a Defect does not have to be corrected	
	Contractor	Reply to the *Contractor's* submission of a proposal to change the Scope so that a Defect does not have to be corrected – acceptance or non-acceptance[b]	Within the *period for reply*
45.2	*Contractor*	Reply to the *Contractor's* submission of a quotation to change the Scope so that a Defect does not have to be corrected – acceptance or non-acceptance	Within the *period for reply*
	Contractor	Instruction to change the Scope, the Prices, the Completion Date and the Key Dates accordingly and to accept the revised programme	
46.1	*Contractor*	Notification that the *Contractor* was given access to correct a notified Defect but the Defect was not corrected within its *defect correction period* and that the *Project Manager* will assess the cost to the *Contractor* of correcting the Defect and include it in the assessment of the amount due as a payment due from the *Contractor*[b]	
46.2	*Contractor*	Notification that the *Contractor* will not be given access to correct a notified Defect and that the *Project Manager* will assess the cost to the *Contractor* of correcting the Defect and include it in the assessment of the amount due as a payment due from the *Contractor*[b]	
51.1	*Contractor* *Client*	Payment certificate	Within 1 week of each assessment date
53.1	*Contractor* *Client*	Payment certificate (final payment)	4 weeks after the *Supervisor* issues the Defects Certificate or 13 weeks after the *Project Manager* issues a termination certificate[a]
61.1	*Contractor*	Notification of a compensation event	At the time of the triggering event
61.2	*Contractor*	Instruction to submit quotations included in the notification of the compensation event with two exceptions	At the time of notifying a compensation event
61.4	*Contractor*	Reply to the *Contractor's* notification of a compensation event Notification that the Prices, Completion Date and the Key Dates are not to be changed Or Notification that the event is a compensation event and includes in the notification an Instruction to submit quotations	Within 1 week of the *Contractor's* notification or longer as agreed
	Contractor	Reply to the *Contractor's* notification of the *Project Manager's* failure to reply on time (Further no response leads to acceptance of the event as a compensation event and a deemed instruction to submit quotations)	Within 2 weeks of the *Contractor's* notification

Clause	To whom	About what	Timing[a]
61.5	*Contractor*	Notification that the *Contractor* did not give an early warning of the event that an experienced contractor could have given – included in the instruction to the *Contractor* to submit quotations	At the same time as the instruction to submit quotations
61.6	*Contractor*	States assumptions in their instruction	At the same time as the instruction to submit quotations
62.1	*Contractor*	Instruction to submit alternative quotations	
62.3	*Contractor*	Reply to the *Contractor's* quotation submission – four choices for the reply	Within 2 weeks of the submission
62.4	*Contractor*	Instruction to submit a revised quotation and explaining reasons for doing so	
62.5	*Contractor*	Informs of the extension that has been agreed for the *Contractor* to submit quotations for a compensation event or for the *Project Manager* to reply to a quotation[b]	
62.6	*Contractor*	Reply to the *Contractor's* notification that the *Project Manager* did not reply to a quotation within the time allowed – original four choices and a comment on the *Contractor's* choice of quotation (Further no response leads to acceptance of the event as a compensation event and a deemed instruction to submit quotations)	Within 2 weeks of *Contractor's* notification
63.2	*Contractor*	Notification of the *Project Manager's* agreement of rates and lump sums to assess the change to the Prices[b]	
63.11	*Contractor*	The *Project Manager* corrects the description of the Condition for a Key Date	
64.3	*Contractor*	Notification of the *Project Manager's* assessment of a compensation event with details	Within the same period allowed for the *Contractor's* submission of the quotation for the same compensation event
64.4	*Contractor*	Reply to the *Contractor's* notification that the *Project Manager* did not assess a compensation event within the time allowed and a comment on the quotation the *Contractor* proposes to be used (Further no response leads to deemed acceptance of the quotation by the *Project Manager*)	Within 2 weeks of the *Contractor's* notification
65.1	*Contractor*	Instruction to submit quotations for a proposed instruction, with the date by which the proposed instruction may be given	
65.2	*Contractor*	Reply to the *Contractor's* quotation for a proposed instruction – three replies described	By the date when the proposed instruction may be given
65.3	*Contractor*	Instruction with notification as a compensation event and instruction to submit a quotation	
70.2	*Contractor*	Permission to remove Plant and Materials from the Working Areas[b]	

Clause	To whom	About what	Timing[a]
72.1	*Contractor*	Notification that Equipment may be left in the *works*[b]	
73.1	*Contractor*	Instruction dealing with objects of value or historical or other interest	
84.1	*Contractor*	Reply to *Contractor's* submission of certificates which state that the insurance required by the contract is in force for acceptance – acceptance or non-acceptance with a reason	Within 2 weeks of submission
86.1	*Contractor*	Submission of certificates for insurance provided by the *Client*	Before the *starting date* and as the *Contractor* instructs
90.1	*Contractor* *Client*	Termination certificate	Promptly
91.2	*Contractor*	Notification that the *Contractor* has defaulted (three types of default described)	
	Contractor *Client*	Notification that the *Contractor* has not put right the default (three types described) within 4 weeks of the date when the *Project Manager* notified the *Contractor* of the default	After 4 weeks from the date when the *Project Manager* notified *Contractor* of the default
91.3	*Contractor*	Notification that the *Contractor* has defaulted (two types of default described)	
	Contractor *Client*	Notification that the *Contractor* has not stopped the default (five types described) within 4 weeks of the date when the *Project Manager* notified the *Contractor* of the default	After 4 weeks from the date when the *Project Manager* notified the *Contractor* of the default
92.2	*Contractor*	Informs that the *Client* no longer requires the Equipment to complete the *works*[b]	
C, D, E, F20.4	*Contractor*	Reply to a submission of forecasts of the Defined Cost for the whole of the *works*[b]	Within the *period for reply*
C, D, E, F26.4	*Contractor*	Reply to *Contractor's* submission of pricing information in each proposed subcontract document for acceptance – acceptance or non-acceptance with a reason	Within the *period for reply*
C, D, E, F50.9	*Contractor*	Reply to a notification that a part of Defined Cost has been finalised – acceptance, need for further material, and notification of errors	Within 13 weeks of the *Contractor's* notification
	Contractor	Reply to *Contractor's* notification with further materials or correction – choice of 3 responses	Within 4 weeks of the *Contractor's* communication
A55.3	*Contractor*	Reply to the *Contractor's* submission of the revised Activity Schedule for acceptance – acceptance or non-acceptance with a reason within the *period for reply*	Within the *period for reply*
B, D60.6	*Contractor*	Instruction to correct a mistake in the Bill of Quantities	
B, D63.15	*Contractor*	Agree a new priced item is compiled in accordance with the *method of measurement*[b]	

Clause	To whom	About what	Timing[a]
A, B63.16	*Contractor*	Agree a new rate for a category of person in the People Rates[b]	
X2.1	*Contractor*	Notification of a compensation event for a change in the law and an instruction to submit quotations	
X4.2	*Contractor*	Reply to the *Contractor's* submission for acceptance of an alternative guarantor who is also owned by the ultimate holding company – acceptance or non-acceptance with reasons	
X10.3	*Contractor*	An early warning notification as soon as they become aware of any matter that could adversely affect the creation or use of the project's Information Model (and other communications required under the early warning procedure in clause 15)	
X10.4	*Contractor*	Reply to a submission for acceptance of a first Information Execution Plan – acceptance or non-acceptance with reasons	Within 2 weeks of submission
	Contractor	Reply to the *Contractor's* notification that the *Project Manager* did not notify acceptance or non-acceptance within the time allowed (Further no response within 1 week leads to deemed acceptance of the quotation by the *Project Manager*)	Within 1 week of *Contractor's* notification
	Contractor	Reply to a submission for acceptance of a revised Information Execution Plan – acceptance or non-acceptance with reasons (and the same notifications for non-response as above)	Within 2 weeks of submission
X13	*Contractor*	Reply to a submission for acceptance of a bank or insurer that will provide a performance bond – acceptance or non-acceptance with reasons	Within the *period for reply*
X14	*Contractor*	Reply to a submission for acceptance of a bank or insurer that will issue an advanced payment bond – acceptance or non-acceptance with reasons	Within the *period for reply*
X16	*Contractor*	Reply to a submission for acceptance of a bank or insurer that will provide a retention bond – acceptance or non-acceptance with reasons	Within the *period for reply*
X21.3	*Contractor*	Reply to the *Contractor's* proposal that the Scope is changed in order to reduce the cost of operating and maintaining an asset. The reply is non-acceptance or acceptance and an instruction to submit a quotation[b]	Within the *period for reply*
	Contractor	Reply to a quotation submitted for acceptance – acceptance or non-acceptance with reasons	Within the *period for reply*
	Contractor	Instruction to change the Scope, the Prices, the Completion Date and the Key Dates accordingly and accept the revised programme	

Clause	To whom	About what	Timing[a]
X22	*Contractor*	Reply to a submission for acceptance of the *Contractor's* detailed forecasts of the total Defined Cost of the work to be done in Stage One – acceptance or non-acceptance with reasons	Within 1 week of submission
	Contractor	Reply to a submission for acceptance of the *Contractor's* forecast of the Project Cost – acceptance or non-acceptance	Within 1 week of submission
	Contractor	Reply to a submission for acceptance of the *Contractor's* design proposals for Stage Two – acceptance or non-acceptance with reasons	Within the *period for* reply or as stated in the Scope
X22.5	*Contractor*	Instruction to replace a *key person* during Stage One	
X22.6	*Contractor*	Notice to proceed to Stage Two Or Instruction to remove from the Scope the work required in Stage Two	
Y1.4	*Contractor*	Reply to a submission for acceptance of the *Contractor's* details of the banking arrangements for the Project Bank Account – acceptance or non-acceptance with a reason	Within the *period for reply*
Y1.6	*Contractor*	Reply to submission for acceptance of the *Contractor's* proposal for adding a Supplier to the Named Suppliers – acceptance or non-acceptance with a reason	Within the *period for reply*

[a] As required unless otherwise stated
[b] Not a communication required by the ECC

NEC4: The Role of the *Project Manager*
ISBN 978-0-7277-6353-2

Appendix 2

Checklist for what the *Project Manager* needs to know from the *Client*

Clause	What the *Project Manager* needs to know from the *Client*	✓
11.2(2)	■ Does the *Project Manager* know how the *Client* is intending to use the *works*? ■ Does the *Project Manager* know what Others need to do their work?	
11.2(8), 13.1	■ Does the *Client* want the Early Warning Register (and other project documents) to be in a specific format or created and updated using specific software that can track documents and is searchable?	
11.2(18)	■ How does the *Project Manager* address any questions that they have about the Site Information or the Site?	
11.2(20), 16.1	■ Has the *Client* accepted the *Contractor's* proposals in their tender for the Working Areas? ■ How does the *Project Manager* make sure that any changes proposed by the *Contractor* are acceptable to the *Client*?	
13.2	■ Does the *Project Manager* understand the operation of and have access to the communications system?	
13.3	■ Does the *period for reply* provide enough time for the *Client* and the *Project Manager* to confer if required? Also ■ What timescale is required? ■ What are the *Client's* restrictions and rules about the *Project Manager* making the decisions they are required to make by the ECC?	
13.5	The *Project Manager* can agree an extension to the *period for reply* on behalf of the *Supervisor* ■ Does the *Client* need to give permission for this to happen? ■ (From the *Supervisor*) What communications should there be between the *Project Manager* and the *Supervisor* to facilitate the change?	
14.2	■ Does the *Project Manager* need to seek permission from the *Client* if they wish to delegate some of their duties? ■ Will the *Client* **require** the *Project Manager* to delegate some of their duties (e.g. core clause section 5)?	
14.3	■ Does the *Project Manager* need to seek permission from the *Client* before they provide an instruction to the *Contractor* that changes the Scope or a Key Date?	
15.1	■ Does the *Client* want communication about any early warnings?	
16.2	■ Does the *Client* want other representatives at the early warning meetings?	
16.3	■ Are there any restrictions on the *Project Manager* with regards to proposals and decisions to make in an early warning meeting?	

Clause	What the *Project Manager* needs to know from the *Client*	✓
16.4	■ Does the *Client* want copies of the Early Warning Register?	
17.1	■ Does the *Client* require to be conferred with before the *Project Manager* gives an instruction changing the Scope?	
18.1	■ Does the *Client* require to be conferred with before the *Project Manager* gives an instruction changing the Scope?	
19.1	■ Does the *Client* require to be conferred with before the *Project Manager* gives an instruction dealing with the event?	
21.2	■ Is the *Project Manager* required to delegate this action or are there any rules regarding the review and acceptance of the design?	
23.1	■ Does the *Client* want to see particulars of the design of an item of Equipment?	
24.1	■ Does the *Client* want to give input into the replacement of key people?	
25.2	■ How does the *Client* want the *Project Manager* to communicate with the *Client* when they should provide services and other things?	
25.3	■ How will the *Client* communicate with the *Project Manager* that it has incurred additional cost?	
26.2	■ Does the *Client* want more control over the choice of Subcontractors? ■ What procedures does the *Client* want the *Project Manager* to follow regarding accepting Subcontractors?	
27.3	■ Does the *Client* want to approve any instructions before the *Project Manager* gives them to the *Contractor*?	
31.3	■ Does the *Client* want to give input before accepting the programme? ■ Does the *Client* want dialogue with the *Project Manager* about the programme and any barriers to Completion?	
33.1	■ How does the *Client* wish the *Project Manager* to communicate forewarning that an *access date* is approaching and that the *Client* is required to provide access?	
34.1	■ What forewarning does the *Client* require prior to the *Project Manager* stopping/not starting/restarting/starting work?	
35.1	■ What communication does the *Client* want from the *Project Manager* before the *Client* takes over the *works*?	
35.2	■ How will the *Project Manager* receive communication that the *Client* plans to use the *works*?	
35.3	■ Is the *Project Manager* required to seek permission from the *Client* before take over is certified?	
36.1	■ Does the *Project Manager* need to consult with the *Client* before they propose an acceleration to achieve Completion before the Completion Date? ■ Does the *Client* want to be consulted if the *Contractor* proposes an acceleration to achieve Completion before the Completion Date?	
36.2	■ Does the *Client* wish to see the *Contractor's* quotation prior to the *Project Manager's* acceptance?	
40.2	■ Does the *Client* or one of their experts want input into the quality policy statement and/or quality plan?	
41.2	■ What forewarning does the *Client* require of the *Project Manager* before providing materials, facilities and samples for tests/inspections?	
41.4	■ Does the *Client* want to know if tests/inspections show a Defect?	

Clause	What the *Project Manager* needs to know from the *Client*	✓
41.5	▪ Does the *Client* want to know if the *Supervisor* causes unnecessary delay to the work or a payment that is conditional upon a test or inspection being successful?	
43.2	▪ Does the *Client* require the *Project Manager* to keep the list of notified Defects and/or to copy it to the *Client*?	
44.3	▪ Does the *Client* want the *Project Manager* to communicate the contents of the Defects Certificate before the *Supervisor* issues it to the *Contractor*?	
44.4	▪ How should the *Project Manager* communicate with the *Client* about access for a part of the *works* that has been taken over?	
45.1, 45.2	▪ Does the *Client* wish to have dialogue with the *Project Manager* before Defect acceptance is proposed and any quotation accepted?	
46.1,46.2	▪ What dialogue does the *Client* expect from the *Project Manager* if a Defect is to remain uncorrected?	
50, 51, 53	The *Project Manager* and the *Client* will need to interact comprehensively throughout the payment procedure, including the final assessment as the *Project Manager* is committing the *Client* to make payments	
60.1(2), (3), (5), (14), (16), (18)	Several compensation events arise because the *Client* has not done something they should ▪ What interactions does the *Client* want with the *Project Manager* to prevent this happening? ▪ How does the *Client* want to deal with compensation events that have arisen due to the *Project Manager's* default?	
66.1	The *Client* is not a part of the compensation event procedure ▪ Does the *Client* wish to be kept informed for each compensation event or sent a monthly update? ▪ Does the *Client* want to be advised prior to a compensation event being implemented under clause 66.1?	
70–73	Title passes from the *Contractor* to the *Client* and back, but the most important transition is where Plant and Materials vest with the *Client* even though they are outside the Working Areas; given that large payments may be made for, for example, modular structures, risk passing may also be required ▪ Does the *Client* want to know about any vesting?	
82	▪ What dialogue does the *Client* want to have with the *Project Manager* about events for which the *Client* is liable, and similarly for events for which the *Contractor* is liable?	
84.1	▪ Does the *Client* wish to be kept informed about which certificates that state that the insurance required by the contract is in force are in place?	
86.1	▪ How will the *Client* provide the required certificates to the *Project Manager*?	
90.1	Communications about termination take place between the *Contractor* and the *Project Manager*, but termination affects the *Client* far more than the *Project Manager* ▪ What procedures and/or communication does the *Client* want the *Project Manager* to undertake with regards to termination using the ECC?	
C, D, E20.3	▪ Does the *Client* want to be part of the discussion when the *Contractor* advises the *Project Manager* on the practical implications of the design of the *works* and on subcontracting arrangements?	
C, D, E, F20.4	▪ Does the *Client* want sight of the forecasts of the total Defined Cost for the whole of the *works*?	
C, D, E, F26.4	▪ Does the *Client* want to see the pricing information in the proposed subcontract documents for each subcontract, to assess the commercial implications?	

Clause	What the *Project Manager* needs to know from the *Client*	✓
A, B63.12	▪ What dialogue does the *Client* want about reduction to the Prices based on the *value engineering percentage*?	
A, B63.16	▪ Does the *Client* wish to be consulted about new rates for a category of person in the People Rates?	
W1, W2, W3	▪ What procedures does the *Client* want the *Project Manager* to follow when a dispute is notified by either Party?	
X2	▪ What communication does the *Client* want to see from the *Project Manager* if there has been a change in the law?	
X4, X13, X14, X20, Y(UK)1	▪ How does the *Client* want this communication to take place?	
X12	▪ Depending on the Partners, Core Group, etc., what interaction does the *Client* require from the *Project Manager*?	
X22	▪ What consultation does the *Client* want from the *Project Manager* with regard to design proposals, the total of the Prices, the notice to proceed to Stage Two, changes to the Budget and the final Project Cost?	
Y(UK)1	▪ What consultation does the *Client* want from the *Project Manager* with regard to banking arrangements, Named Suppliers and the Trust Deed?	

NEC4: The Role of the *Project Manager*
ISBN 978-0-7277-6353-2

ICE Publishing: All rights reserved
http://dx.doi.org/10.1680/nectrpm.63532.103

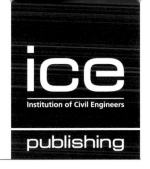

Appendix 3
Checklist for what the *Project Manager* can look for in the contract documentation

	Things to look for in the contract documentation	✓
1	Dates to be marked on the *Project Manager's* wallchart: ■ *starting date* ■ *access dates* ■ Conditions and Key Dates ■ Completion Date ■ planned Completion ■ assessment of the amount due ■ critical path dates ■ dates for critical tests/inspections ■ induction of critical Subcontractors ■ other dates that are required by clause 31.2 and that the *Project Manager* has notice of ■ dates when the *Client* is required to provide things or do something ■ early warning meetings ■ dates for regular meetings with other people involved in the project ■ potential dates for inspections by statutory bodies	
2	Factors that may affect the Prices and date of Completion, such as ■ winter working ■ critical work that has been subcontracted ■ specialist tests/inspections	
3	Accepted Programme – check information in clause 31.2, such as ■ key programme information ■ the order and timing of the operations of the *Client* and Others ■ the statement of how the *Contractor* plans to do the work, identifying the principal Equipment and other resources that they plan to use	
4	Activities for which the *Project Manager* wants to be present, such as ■ a specialist Subcontractor's first day on the Site ■ installation of key M&E kit ■ test/inspection on the critical path or for which other people have to be present ■ final activities before sectional Completion/take over ■ high-risk activities, such as dropping an escalator into the building	
5	Additional *Client* liabilities that may affect the way in which some activities are approached	
6	Option C or D: the *Contractor's share range* and *share percentage*, especially if frequent and/or large compensation events are implemented	
7	Interaction of any *sections* (with Options X6 and X7)	
8	Option X12 – partnering (not used with Option X20): ■ check the Schedule of Partners ■ read the Partnering Information ■ note any Key Performance Indicators	

	Things to look for in the contract documentation	✓
9	Use of Option X20 (not used with Option X12): what are the proposed Key Performance Indicators?	
10	Use of Option X22: dates for forecasts, design submissions, and Stage Two dates and requirements	
11	What does the *Contractor's* data for the SCC look like (for Option C, D and E contracts)?	
12	What does the *Contractor's* data for the SSCC look like and are there any possible compensation events that will skew the Prices?	
13	Are there any areas of the Scope that the *Project Manager* thinks need to be more carefully supervised?	
14	Overall, is there anything about the *Contractor's* tender that looks odd or out of place, especially considered against the Scope and the Site/Working Areas?	
15	Are there any aspects of the Site that might cause problems? For example ■ its location (noise, lights, health and safety, air ownership) ■ underground unknowns ■ access ■ security	
16	Are the Working Areas likely to cause any problems through their location or size?	
17	What is already included on the Early Warning Register?	

NEC4: The Role of the *Project Manager*
ISBN 978-0-7277-6353-2

ICE Publishing: All rights reserved
http://dx.doi.org/10.1680/nectrpm.63532.105

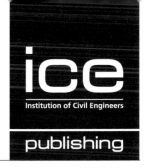

Appendix 4
Checklist for examples of things to look for in the Scope and Contract Data

Clause	Things to look for in the Scope and Contract Data	✓
11.2(2), 30.2	■ Does the Scope list all the work that the *Contractor* must do before the Completion Date? ■ Does the Scope detail what the *Client* needs to be completed before they can use the *works*? ■ Does the Scope detail what Others need to do their work?	
11.2(6)	■ Does the Scope provide enough detail so that a Defect is easily discernible?	
11.2(16)	■ Does the Scope specify and describe the *works* and state constraints on how the *Contractor* Provides the Works?	
13.2	■ Does the Scope specify the use of a communication system (e.g. software or a numbering system), and how does this affect the *Project Manager*?	
13.3 with CD1	■ What difficulties might the *period for reply* bring? Does CD1 provide different periods for different communications/scenarios? ■ How will the differences impede/enhance the contract?	
15.1 and 15.2 with CD1 and CD2	■ Is the Early Warning Register already in existence, using the information provided in the Contract Data by both Parties?	
17.2	■ Does the Scope include an illegal or impossible requirement?	
21.1	■ Does the Scope require the *Contractor* to undertake any design of the *works*?	
21.1	■ What particulars of design does the Scope require the *Contractor* to submit to the *Project Manager*?	
22.1	■ Does the Scope state something other than that the *Client* may use and copy the *Contractor's* design? ■ Does the Scope provide other purposes for which the *Client* may use the *Contractor's* design?	
25.1	■ Does the Scope provide details about how the *Contractor* should co-operate with Others and share the Working Areas with them?	
25.2	■ Does the Scope provide details about services and other things that must be provided by the *Client* and the *Contractor*?	
27.4	■ What health and safety requirements are included in the Scope?	

31.2	■ Does the Scope include
	  – the order and timing of the work of the *Client* and Others?
	  – any procedures that the *Client* wants the *Contractor* to adhere to?
	  – information that the *Client* requires the *Contractor* to show on the programmes that they submit for acceptance?
	  – the form in which the programme must be issued for acceptance?
	■ Does the Contract Data describe the *conditions* and *key dates*?	

Clause	Things to look for in the Scope and Contract Data	✓
31.3	■ What does the Scope require the programme to look like, include or dictate?	
32.2	■ How often is the *Project Manager* to receive revised programmes? ■ What impact will this have on the *Project Manager's* time and other activities?	
35.2	■ What reason does the Scope provide for the *Client* to use any part of the *works* before Completion has been certified?	
40.1	■ What requirements for operating a quality management system are included in the Scope?	
41.1	■ What tests/inspections are required by the Scope?	
41.2	■ What materials, facilities and samples does the *Client* need to provide for the tests/inspections?	
42.1	■ Is a list of Plant and Materials that the Scope requires to be tested/inspected before delivery provided?	
44.2	■ Does CD1 provide different defect correction periods? ■ How do these impact on the *Project Manager's* activities or the project?	
50.2	■ Does the Scope describe the form in which the *Contractor* must submit applications for payment?	
60.1(5)	■ Does the Scope describe the conditions within which the *Client* or Others must work? ■ Does the Scope describe the work that must be carried out by the *Client* and Others on the Site?	
60.1	■ Does the Contract Data detail other compensation events?	
71.1	■ Does the Scope describe how the *Contractor* is to prepare for marking Equipment, Plant and Materials that are outside the Working Areas? ■ Does the Scope identify for payment Equipment, Plant and Materials that are outside the Working Areas?	
73.2	■ Does the Scope state what title the *Contractor* has to materials from excavation and demolition?	
80.1	■ How will the *Project Manager* deal with any additional liabilities in CD1?	
83	■ Does the Contract Data describe any additional insurances?	
C, D, E11.2(26), F11.2(27)	■ Does the Scope describe acceptance or procurement procedures?	
C, D, E, F52.2	■ Does the Scope describe the records that the *Contractor* is required to keep (related to payments)?	
A, C55.1	■ Do the activities in the Activity Schedule relate to the Scope?	
B, D56.1	■ Does the Scope include all the Scope-relevant descriptions that would normally appear in a Bill of Quantities?	
X4.1	■ Does the Scope include the form of the ultimate holding company guarantee?	
X8.4	■ Does the Scope include the form for the *undertaking to Others*, the *Subcontractor undertaking to Others* and the *Subcontractor undertaking to the Client*?	
X9.1	■ Does the Scope describe ownership other than the *Client's* ownership of the *Contractor's* rights over material prepared for the design of the *works*? ■ Does the Scope describe other rights for the *Client* that the *Contractor* must obtain?	

Clause	Things to look for in the Scope and Contract Data	✓
X10.1(4)	■ Does the Scope describe the requirements for creating the Information Model?	
X13.1	■ Does the Scope include the form of the performance bond?	
X14.2	■ Does the Scope include the form of the advanced payment bond?	
X15.3	■ Does the Scope describe a situation other than that the *Contractor* may use the material provided by it under the contract for other work?	
X15.4	■ Does the Scope describe the form in which copies of drawings, specification, records and other documents recording the *Contractor's* design are to be retained?	
X16.3	■ Does the Scope include the form of the retention bond?	
X22.1	■ Does the Scope describe Stage One and Stage Two?	
X22.2	■ Does the Scope describe forecasts, what is to be included and how they are to be presented?	
X22.3	■ Does the Scope describe the submission procedures for the *Contractor* submitting design proposals for Stage Two to the *Project Manager*?	
X22.3(6) and X22.6(1)	■ Does the Scope describe the approvals and consents required from Others?	
X22.6(3)	■ Does the Scope provide the performance requirements?	
Y1.6	■ Does the Scope provide information about adding suppliers?	
General	■ Does the Scope add to existing ECC procedures or create new procedures that need to be followed by the *Project Manager*? For example – clause 15.1: additional procedures for managing the Early Warning Register – clause 15.2: procedures for conducting and attending an early warning meeting – clause 21.2: the *Contractor's* submission of the particulars of their design – clause 24.1: other procedures for the replacement of key people	

NEC4: The Role of the *Project Manager*
ISBN 978-0-7277-6353-2

ICE Publishing: All rights reserved
http://dx.doi.org/10.1680/nectrpm.63532.109

Appendix 5
Agenda for the first meeting between the *Project Manager* and the *Supervisor*

1. Communications
   - Form of communication
   - Procedures for the *period for reply*
   - Document format and recipients
   - Delegation procedures
   - Early warning procedures
   - Review of and amendments to contract documents
   - Communication about day-to-day activities
   - Meetings with the *Contractor*

2. Regular meetings and agendas
   - Meetings about programme matters and the amount due
   - Early warning meeting
   - Meeting about Defects
   - Meeting to discuss possible compensation events

3. Quality management system
   - Checks against the *Contractor's* quality management system
   - Management using the *Contractor's* quality plan

4. Procedures regarding tests and inspections
   - Tests/inspections on which a payment is conditional

5. Procedures about Defects
   - Requirements for clause 44.2 – correcting Defects before Completion
   - Procedure for extending/changing the *defect correction period*
   - Procedure to instruct an immediate correction of a Defect
   - Procedure for searches
   - Procedures about the acceptance of Defects
   - Defects affecting payment, the programme and the Scope

6. Procedures about marking Plant and Materials
   - The *Contractor's* requirements
   - Recording of information and records
   - Site diaries
   - Weather records
   - Records of activities in the Working Areas

NEC4: The Role of the *Project Manager*
ISBN 978-0-7277-6353-2

ICE Publishing: All rights reserved
http://dx.doi.org/10.1680/nectrpm.63532.111

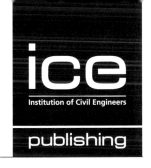

Appendix 6
Checklist for the kick-off meeting with the *Contractor*

Clause	Description	Activity (Establish the working practices that will enable the contract to be operated correctly)
General		
13	Communication protocol	■ Consider any communication system included in the Scope and the ramifications of how the *Project Manager* and the *Contractor* will communicate ■ Establish the protocol, templates, use of software and the cloud system ■ Confirm the contact address for notices
15	Early warnings	■ Issue the Early Warning Register ■ Establish the timing for regular early warning meetings
16.3	Working Areas	■ Identify and confirm the Working Areas
17.1	Ambiguities and inconsistencies	■ Discuss any issues that may have come to light
17.2	Illegal or impossible requirements	■ Discuss any issues that may have come to light
20	The *Contractor's* main responsibilities	■ Discuss any issues specific to the project, e.g. decanting of occupants of the building ■ Discuss any parts of the Scope that need emphasising or on which the *Contractor* has queries
21	The *Contractor's* design submissions for the *works*	■ Discuss the process for design submissions for the *works*
23	The *Contractor's* design submissions for Equipment	■ Discuss the process for design submissions for Equipment
24	Key persons	■ Check that key people named in CD2 are on the project
25	Others	■ Discuss what Others will be working in the Working Areas and the possible impact on the *works*
26	Subcontracting	■ Submission of the names of the *Contractor's* proposed Subcontractors if not already established at the tender stage ■ Submission of the proposed subcontract documents ■ Introduction to the Subcontractors' key people
25, 27	Working with Others	■ Possible impact on the *works* of Others in the Working Areas ■ Establishment of communications lines with Others

Clause	Description	Activity (Establish the working practices that will enable the contract to be operated correctly)
Options C, D, E and F		
20.3	Practical implications of the design of the *works* and on subcontracting arrangements	▪ Establish the process
20.4	Forecast of total Defined Cost for the whole of the *works*	▪ Establish consultation process
Time		
30	Generally, check key programme dates	▪ *starting* date ▪ *access dates* ▪ planned Completion ▪ Completion Date ▪ Completion Dates for *sections* of the *works* ▪ Key Dates
31.1	First programme submitted for acceptance	▪ Work with the *Contractor* to ensure that their first programme is submitted for acceptance within the timescales required by CD1
32.2	Monthly submission of the programme	▪ Establish the monthly working practice for revisions
Testing and Defects (mainly the actions of the *Supervisor* and the *Contractor*)		
40	Quality management system	▪ Are the quality policy statement and quality plan on record? ▪ How is the *Contractor* going to make sure that they adhere to a quality management system that complies with the requirements in the Scope?
41.1, 41.3	Tests and inspections	▪ Have the *Supervisor* and the *Contractor* established a plan?
42.1	Tests and inspections before delivery to the Working Areas	▪ Have the *Supervisor* and the *Contractor* established a plan?
43.2	Defects notification	▪ Have the *Supervisor* and the *Contractor* established the Defects notification process?
Payment		
50.1	First assessment date	▪ Establish the first assessment date to suit the Parties
50.2, 50.3	Amount due	▪ Set up processes for the submission of the application for payment and to determine the amount due
51.1	Payment certificate	▪ Set up a payment certificate
51.2	Certify payment	▪ Establish the payment protocol with the *Client*
52.1	Defined Cost	▪ Check to see that the *Contractor* is operating Defined Cost properly
Option A		
11.2(29)	Price for Work Done to Date	▪ Completed activities

Clause	Description	Activity (Establish the working practices that will enable the contract to be operated correctly)
Option B		
11.2(30)	Price for Work Done to Date	■ Quantity of work completed
Option C, D, E and F		
11.2(31)	Price for Work Done to Date	■ Defined Cost
Compensation events		
6	Compensation events	■ Implement core clause section 6: notify, quote, assess and implement ■ Establish the discussion process for quotations ■ Establish the meeting routine for the management of compensation events ■ Option B and D choices
Options B and D		
6	Bill of Quantities	■ Process for compensation events arising due to the use of the Bill of Quantities
Title (mainly the action of the *Supervisor*)		
70	Title to Plant and Materials	■ Discuss the procedures for Plant and Materials on the Site and/or within the Working Areas
71.1	Marking of Equipment, Plant or Materials outside the Working Areas for payment	■ Marking and vesting requirements as stated in the Scope
72.1	Removing Equipment from the Site	■ Establish procedures
Risk and Insurances		
84.1	*Contractor's* insurances	■ Have the insurances been submitted?
86.1	*Client's* insurances	■ Have the insurances been submitted?
Termination		
		No initial actions required

NEC4: The Role of the *Project Manager*
ISBN 978-0-7277-6353-2

ICE Publishing: All rights reserved
http://dx.doi.org/10.1680/nectrpm.63532.115

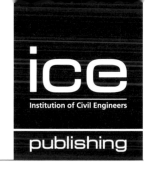

Checklist for templates that can be used by the *Project Manager*

Clause	Description	Template type
14.2	Delegation	Notification
	Cancel delegation	Notification
14.3	Change the Scope or Key Date	Instruction
15.1	Early warning	Notification
15.2	Attend an early warning meeting	Instruction
15.1, 15.4	Issuing the Early Warning Register	Issue
15.4, 16.2	Change to the Scope	Instruction
16.2	Proposal that the Scope is changed to reduce the total of the Prices – reply	Proposal reply
	Contractor to submit quotations	Instruction
16.3	Proposal to add to Working Areas – reply	Proposal reply
17.1	Ambiguity or inconsistency	Notification
	Resolve the ambiguity or inconsistency	Statement
17.2	The Scope includes an illegal or impossible requirement	Notification
	Change to the Scope	Instruction
19.1	Deal with an event	Instruction
21.2	Submission of the particulars of design – for acceptance	Reply
23.1	Submit the particulars of design of an item of Equipment	Instruction
	Design of an item of Equipment – for acceptance	Reply
24.1	Submit a proposed replacement person – for acceptance	Reply
24.2	Remove a person	Instruction
25.3	Work has not met the Condition for a Key Date on time	Notification (not required by the ECC)
26.2	Name of a proposed Subcontractor – for acceptance	Reply

Clause	Description	Template type
26.3	Proposed subcontract documents – for acceptance	Reply
C, D, E, F26.4	Pricing information for proposed subcontract documents – for acceptance	Reply
27.2	Provide access to Others	Instruction
30.2	Completion certificate	Certificate
31.3	Programme submission – for acceptance	Reply
34.1	Stop or not start work/restart or start work	Instruction
35.3	Take over certificate	Certificate
36.1	Provide a quotation for acceleration	Instruction
	Submit a quotation for acceleration – for acceptance	Reply
	Proposal to accelerate	Proposal
36.3	Change to the Prices, the Completion Date and the Key Dates; and accept the revised programme	Instruction and acceptance
40.2	Submit the quality policy statement and quality plan – for acceptance	Reply
40.3	Correct a failure to comply with the quality plan	Instruction
45.1	Change the Scope to accept a Defect	Proposal
	Change the Scope to accept a Defect – reply	Reply
45.2	Submit a quotation for reduced Prices or an earlier Completion Date or both – for acceptance	Reply
	Change to the Scope, Prices, Completion Date and Key Dates; and accept the revised programme	Instruction and acceptance
51.1 and 53.1	Payment certificate	Certificate
A55.3	Submit a revision of the Activity Schedule – for acceptance	Reply
61.1, X2.1 …	Notification of a compensation event (see list in section 6.5 of this book)	Notification
61.2	Submit quotations	Instruction
61.4	Decision about a compensation event notification	Reply
61.5	Failure to notify an early warning	Notification
62.1	Submit alternative quotations	Instruction
62.3	Quotation submission – reply	Reply
62.4	Revised quotations	Instruction
62.6	*Project Manager* failed to respond	Notification
64.3	*Project Manager's* assessment	Notification
65.1	Submit a quotation for a proposed instruction	Notification
65.2	Quotation submitted – for acceptance	Reply

Clause	Description	Template type
65.3	Issue an instruction	Instruction
	Instruction is a compensation event	Notification
	Submit a quotation	Instruction
66.1	Acceptance of a quotation	Notification
	Assessment made by the *Project Manager*	Notification
70.2	Removal of items from the Working Areas	Permission
72.1	Equipment left in the *works*	Permission
73.1	How to deal with an object of value or historical or other interest	Instruction
84.1	Certificate submission – for acceptance	Reply
86.1	Submission of policies and certificates	Submission
90.1	Termination certificate	Certificate
91.2, 91.3	Notification of *Contractor* default with no correction	Notification
Y1.4	Submission – reply	Reply
Y1.6	Submission – reply	Reply

NEC4: The Role of the *Project Manager*
ISBN 978-0-7277-6353-2

Appendix 8
Regular meeting agendas

Appendix 8A: Agenda for the weekly meeting

Not all aspects of the agenda will need to be covered in every project and every week. The *Project Manager* should choose those topics that they want to discuss in their weekly meeting and issue an agenda that covers those topics.

1. Introduction and welcome

2. Minutes of the previous meeting

3. Report from the designers

4. Plant and Materials
   - Site fabrication and delivery

5. Report from the *Supervisor*
   - General quality matters
   - Tests/inspections
   - Defects
   - Plant and Materials ready to be moved into the Working Areas
   - Vesting

6. Synopsis of instructions issued in the previous week

7. Report on the Accepted Programme
   - Progress over the past week and projected progress for the next week
   - Any early warnings
   - Notifications, and permissions needed from Others to progress

8. Compensation events
   - Submissions in the past week
   - Progress on those being assessed
   - Compensation event matters projected for the next week

9. Health and safety and environmental issues

10. Advance warnings of any regulatory inspections of the Site

11. Drawings and operations and maintenance manuals

12. Special mentions this week
    - For example, prefabricated modules arriving and what set-up is required for them

13. Any other business

14. Date of the next meeting

Appendix 8B: Agenda for the Monthly meeting with the project team, including the *Contractor* and Others

1. Introduction and welcome
2. Minutes of the previous meeting
3. Design
4. Working Areas
   - Additions or changes and the resulting impact on Defined Cost
5. People issues
6. Instructions
7. Accepted Programme
   - Progress
   - Access issues
   - Resource levels
   - Look ahead
8. Deliveries of Materials
9. Movement of Equipment, Plant and Materials on and off the Site and Working Areas
10. Subcontractors
    - Proposals of new/replacement Subcontractors
    - Submission of subcontract documents under clause 26
    - Any issues (e.g. a replacement person)
11. Commercial and procurement
    - Update on Prices
    - Payment
      - Statement on the previous payment certificate
      - Any disputed amounts
      - Comments on the previous application for payment and any changes to be made in presentation
      - Defined Cost
        ○ Forecast of Defined Cost for the whole of the *works*
    - Procurement of Subcontractors
    - Notification of forthcoming audit
12. Quality matters
    - Quality of the *works*
      - Compliance with the Scope
      - Compliance with constraints
      - Quality standards
      - Concessions (acceptance of Defects)
      - Workmanship
    - Plant and Materials
    - Compliance with the quality management system, including the quality policy statement and the quality plan
    - Notified Defects
    - Design management
      - Design of the *works*
      - Design of Equipment
13. Early warnings
14. Compensation events
    - Progress
    - Updates
15. Early Warning Register matters
16. Health and safety and the environment
    - Any health and safety issues
    - Any environmental issues
17. Public relations
18. Key Performance Indicators
19. Any other business
20. Date of the next meeting

Appendix 9
Quality management considerations

Clause	Project Manager	Supervisor
40	Be aware of the *Contractor's* quality management system. Accept the quality policy statement and quality plan. Understand the quality plan to check compliance.	Be aware of how the *Contractor's* quality documentation interacts with the Scope for the purposes of notifying Defects and managing tests and inspections.
41.1	Be aware of the tests/inspections included in the Scope and required by law.	Have detailed knowledge of the tests/inspections included in the Scope and required by law.
41.2	Is it the *Project Manager's* or the *Supervisor's* (personal contract) duty to communicate to the *Client* that materials, facilities and samples are required imminently, or are they required to ensure themselves that the materials, facilities and samples are available at the relevant time? Does this determine who must be familiar enough with the materials, facilities and samples stated in the Scope to facilitate their timeous availability?	
41.3	Does the *Project Manager* want the *Supervisor* to report to them regarding any tests/inspections or any results?	Notify tests/inspections and results.
41.4	No duties other than those in 41.3.	No duties other than those in 41.3.
41.5	Does the *Project Manager* want the *Supervisor* to report to them when the test/inspection has been carried out, so that they are able to assess payment at the right time?	No unnecessary delay.
41.6	Assess the cost incurred by the *Client* of repeating a test/inspection after a Defect is found. Does the *Project Manager* want the *Supervisor* to advise them about what is involved so that their assessment is accurate?	No duties.
C, D, E41.7	Consider the *Client's* costs only.	No duties.
42.1	Does the *Project Manager* want the *Supervisor* to notify them of the same thing so that they are aware of any programme issues?	The *Supervisor* to notify the *Contractor* that Plant and Materials have passed the test/inspection.
43.1	Does the *Project Manager* want the *Supervisor* to advise them of any search, what it entails (e.g. additional tests/inspections) and the result so that they can assess any cost or programme implications?	Instruct a search.
43.2	Does the *Project Manager* want the *Supervisor* to provide them with a list/report weekly/monthly?	Notify Defects and receive notifications.

Clause	Project Manager	Supervisor
44.1	No duties.	No duties.
44.2	No duties.	Decide whether to incorporate an indicative *defect correction period* on the Defect notification (may be required if it is a part of the template notification).
44.3	Receive a copy (clause 13.6).	Issue a Defects Certificate.
44.4	Arrange for the *Client* to allow access.	No duties.
45.1	Propose or receive a proposal to change the Scope so that a Defect does not have to be corrected. Does the *Project Manager* want the *Supervisor* to contribute their opinion?	No duties.
45.2	Receive a quotation. If accepted, give an instruction to change the Scope, the Prices and the Completion Date. Does the *Project Manager* want the *Supervisor* to contribute their opinion on how the project is affected by accepting the Defect?	No duties.
46.1	How will the *Project Manager* know that the Defect was not corrected in its *defect correction period*? Assess the cost to the *Client* of having the Defect corrected by other people.	No duties.
46.2	Assess the cost to the *Contractor* of correcting the Defect.	No duties.
70.2	Give permission to remove Plant and Materials from the Working Areas.	No duties.
71.1	Does the *Project Manager* want the *Supervisor* to report when marking has taken place, as it affects the title (clause 70.1), the assessed amount due and the programme?	Mark Equipment, Plant and Materials outside the Working Areas.

NEC4: The Role of the *Project Manager*
ISBN 978-0-7277-6353-2

ICE Publishing: All rights reserved
http://dx.doi.org/10.1680/nectrpm.63532.123

Appendix 10
Checklist for the *Project Manager* for managing quality

Clause	Quality management issue for *Project Manager*	✓
40.2	Has the *Contractor* provided the *Project Manager* with a quality policy statement and a quality plan for acceptance within the period stated in the Contract Data?	
	Is the quality policy statement and the quality plan part of the quality management system described in the Scope?	
40.3	How will the *Project Manager* monitor the *Contractor's* compliance with their quality plan?	
41.1	Is a list of required tests/inspections available?	
41.2	Is it known which tests/inspections require the *Client* to provide materials, facilities and samples?	
41.3	Does the *Project Manager* want notification of tests/inspections before they take place and/or afterwards of the results?	
41.4	Does the *Project Manager* want notification if a test/inspection has revealed a Defect?	
41.5	Does the *Project Manager* want to know if a test upon which a payment is conditional has taken place?	
41.6	Does the *Project Manager* want the *Supervisor's* help with assessing the cost incurred by the *Client* in repeating a test/inspection?	
42.1	Does the *Project Manager* want the *Supervisor* to notify them when tests/inspections required on Plant and Materials have been passed?	
43.1	Does the *Project Manager* want the *Supervisor* to notify them when they have given an instruction to search for a Defect and afterwards of the result?	
43.2	Does the *Project Manager* want the *Supervisor* to notify them of all Defects or just those that would impinge on the Accepted Programme and/or the Prices, and do they want them included automatically on the Early Warning Register?	
44.2	Does the *Project Manager* want the *Supervisor* to notify them for each Defect or when they think the *defect correction period* starts?	
44.4	Does the *Project Manager* want to discuss with the *Supervisor* Defects that require access after take over?	
45.1	Does the *Project Manager* want to discuss with the *Supervisor* proposals to accept Defects?	
46.1	Does the *Project Manager* want the *Supervisor* to notify them if a Defect has not been corrected within its *defect correction period*?	

NEC4: The Role of the *Project Manager*
ISBN 978-0-7277-6353-2

ICE Publishing: All rights reserved
http://dx.doi.org/10.1680/nectrpm.63532.125

Agenda for a meeting about Defects and other matters about quality

1. List of Defects
   - Collated list of Defects
     - Notified by the *Supervisor*
     - Notified by the *Contractor*
     - Notified after a search
     - Result of a failed test/inspection
     - Required correction date (would have prevented the *Client* using the *works* or prevented Others from doing their work)
     - Actual correction date
     - *defect correction period* and where amended/lengthened by agreement
     - Notified after take over
     - Accepted
   - Identified patterns of Defects to be addressed
   - Categories of Defects – is it working?
   - Comment on Defects notified after the failure of a test/inspection
     - Effect on programme
     - Payment conditional on the passing of a test/inspection
     - Assessment of the cost of repeating a test/inspection
   - List of Defects that will be accepted
   - List of instructed searches for Defects

2. Comments about the *defect correction period*
   - List of Defects for which the *defect correction period* was amended (e.g. lengthened)
   - List of Defects for which the *Supervisor* has required immediate correction
   - Is the idea of the *defect correction period* beginning at Completion working?
   - Are fewer/more categories needed?
   - Are changes required?

3. Programme
   - Any Defects that are preventing the completion of an activity
   - Any Defects that are affecting Disallowed Cost

4. Payment
   - Tests or inspections upon which payments are conditional
     - Any such tests/inspections that took place in the previous month that the *Project Manager* will have to take into account
     - Any such tests/inspections that are due in the forthcoming month

5. Accepting Defects
   - List of accepted Defects
   - Defects that the *Supervisor* thinks the *Project Manager* could propose acceptance to the *Contractor*

6. Searches
   - List of searches instructed and carried out and any resulting Defects
   - How the Defects were dealt with (accepted/corrected/changes to the Scope)
   - Any patterns/trends with regards to searches
   - Impact on the programme
7. Quality management system
   - Does the Scope include requirements for a quality management system?
   - Did the *Contractor* provide a quality policy statement and a quality plan to the *Project Manager* for acceptance within the required time period?
   - Has the quality plan changed?
   - Is the *Supervisor* going to notify the *Project Manager* if they think the *Contractor* has failed to comply with the quality plan? If not, how is the *Project Manager* going to know that the *Contractor* has failed to comply with the quality plan?

NEC4: The Role of the *Project Manager*
ISBN 978-0-7277-6353-2

ICE Publishing: All rights reserved
http://dx.doi.org/10.1680/nectrpm.63532.127

Appendix 12
Early warning meeting agendas

Appendix 12A: Agenda for a special early warning meeting

1. Reason for the early warning meeting
   - Description of the early warning notification
   - What prompted the notification?
   - Who gave the notification?
   - What timescales are involved?

2. Description of the matter for incorporation on the Early Warning Register

3. Proposals for how the effects of the matter can be avoided or reduced

4. Seeking solutions that will bring advantage to those who will be affected

5. Deciding on actions that will be taken and who, in accordance with the contract, will take them

6. Making sure that any Scope changes are clear

7. Date of the next meeting about the matter if different from the next regular early warning meeting

Appendix 12B: Agenda for a monthly early warning meeting

When
Once a month as part of the monthly meeting

Purpose
To discuss the matters on the Early Warning Register

Present
Project Manager *Contractor* *Supervisor* Any specialists affected by any of the risks on the Early Warning Register (e.g. the M&E Subcontractor) – these may change depending on the risks to be discussed.

Agenda

1. Previous minutes
   - Make sure that everyone has the latest copy of the Early Warning Register (the *Project Manager* should have distributed the latest Early Warning Register after the previous early warning meeting, whether this was held as part of the monthly meeting or was called separately after the monthly meeting under clause 15.2)

2. Current matters
   - Discuss the current matters on the Early Warning Register one by one or only those with a higher risk allocation, if using a standard impact/probability chart/matrix (as decided by the team)
   - Assess how the actions required by the previous meeting affected the matter to which they are related, and adjust the status of the matter accordingly
   - Make and consider proposals for how the effect of the registered matters can be avoided or reduced (or agree that circumstances have not changed and no new proposals are viable)
   - Seek solutions that will bring advantage to those who are affected
   - Decide what actions are to be taken and who, in accordance with the contract, will take them
   - Decide which risks have been avoided or passed and can be removed from the Early Warning Register

3. New matters
   - Discuss new matters that have arisen since the previous early warning meeting
   - If a standard impact/probability chart or matrix is being used, those present could agree on the rating to be input
   - Make and consider proposals for how the effect of new matters can be avoided or reduced
   - Seek solutions that will bring advantage to those who are affected
   - Decide what actions are to be taken and who, in accordance with the contract, will take them

4. Close
   - The *Project Manager* summarises
   - The *Project Manager* provides a date by which they will issue the following
     - Minutes relating to the actions decided upon
     - Revised Early Warning Register (the *Project Manager* is required to instruct a change to the Scope at the same time as issuing the revised Early Warning Register to the *Contractor* if a decision made at the early warning meeting was to change the Scope); the revision of the Early Warning Register by the *Project Manager* is deemed to be a record of decisions made at the early warning meeting

NEC4: The Role of the *Project Manager*
ISBN 978-0-7277-6353-2

ICE Publishing: All rights reserved
http://dx.doi.org/10.1680/nectrpm.63532.129

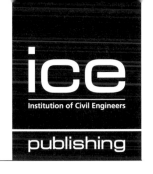

Checklist for sources of information to be included in the first programme

	Information required to be included on the first programme submitted for acceptance by the *Contractor*	Where the *Project Manager* will find the information
1	*starting date*	CD1 – data by the *Client*, under section 3 ('Time').
2	*access dates* for parts of the Site	CD1 – data by the *Client*, under section 3. ■ How many parts of the Site are listed for access?
3	Key Dates	CD1 – data by the *Client*, under section 3, or in the optional statements after section 9.
4	Completion Date	Either in CD1 – data by the *Client*, or in CD2 – data by the *Contractor*, in the optional statements.
5	Date of the planned Completion	Not required to be documented except in the Accepted Programme, so this may be the first time that the *Project Manager* has had sight of the date when the *Contractor* plans to complete the *works*. The date may be before the *completion date* or on the *completion date*.
6	Order and timing of the operations that the *Contractor* plans to do in order to Provide the Works	The Scope may provide constraints about the order and timing of when and how the *works* must be carried out.
7	Order and timing of the work of the *Client* and Others as last agreed with them by the *Contractor* or, if not so agreed, as stated in the Scope	The Scope may provide constraints about the order and timing of the work to be done by the *Client* or Others. As this is the first programme, it is unlikely that arrangements have been made post-contract, and so the *Project Manager* should find what they need in the Scope.
8	Dates that the *Contractor* plans to meet each Condition stated for the Key Dates and to complete other work needed to allow the *Client* and Others to do their work	Key Dates and Conditions to be met by those dates can be found in CD1 by the *Client*. The Scope may describe work to be done by the *Client* and Others and it may also describe the work that needs to be done by the *Contractor* in preparation for the work to be done by the *Client* and Others. 'Others' refers to other contractors and official bodies, such as an environmental/health and safety body or an operational body.

	Information required to be included on the first programme submitted for acceptance by the *Contractor*	Where the *Project Manager* will find the information
9	Provisions for float, time risk allowances, health and safety requirements and the procedures set out in this contract	Float and time risk allowances are not required to exist; it is the *Contractor* who decides whether to include these allowances in the programme they submit for acceptance.
		Health and safety requirements will be included in the Scope, and there may also be health and safety requirements that are described in the law of the contract and which the *Project Manager* needs to be aware of.
		There may be procedures described in the ECC and also in the Scope. Procedures may refer to those set out in the clauses of the ECC, e.g. design acceptances (clause 21.2), acceptance of Subcontractors (clauses 26.2 and 26.3), informing the *Supervisor* of tests/inspections (clause 41.3), marking (clause 71.1); and to those referred to in the clauses of the ECC, e.g. acceptance and procurement procedures (clauses C, D and E11.2(26) and F11.2(27)) and marking procedures (clause 71.1).
10	Dates when, in order to Provide the Works in accordance with their programme, the *Contractor* will need access to a part of the Site if later than its *access date*, acceptances, Plant and Material and other things to be provided by the *Client* and information from Others	*Access dates* can be found in CD1 – data provided by the *Client*, under section 3.
		The requirements for acceptances can be found in the ECC *conditions of contract*, and some may be included in the Scope.
		Plant and Materials and other things to be provided from the *Client* and information from Others will be found in the Scope.
11	For each operation, a statement of how the *Contractor* plans to do the work, identifying the principal Equipment and other resources that they plan to use	This information is part of the *Contractor's* offer. The *Project Manager* should check for any constraints that have been provided in the Scope, e.g. height restrictions for lifting equipment, noise restrictions and space issues.
12	Other information that the Scope requires the *Contractor* to show on a programme submitted for acceptance	This information can be found in the Scope. It may not be in a single place in the Scope; therefore, a complete check needs to be carried out.

NEC4: The Role of the *Project Manager*
ISBN 978-0-7277-6353-2

ICE Publishing: All rights reserved
http://dx.doi.org/10.1680/nectrpm.63532.131

Appendix 14
Checklist for information to be included in the first programme

	Information to be included in the first programme	✓
1	*starting date* ▪ Is the *starting date* provided in CD1 accurately reflected in the programme?	
2	*access dates* ▪ Does the programme accurately reflect the *access dates* for all parts of the Site described in CD1?	
3	Key Dates ▪ Does the programme show the Key Dates provided in CD1?	
4	Completion Date ▪ Is the *completion date* for the whole of the *works* provided as in CD1 or CD2?	
5	date of planned Completion ▪ Has the *Contractor* provided in the programme a date for planned Completion?	
6	Order and timing of the operations that the *Contractor* plans to do in order to Provide the Works ▪ Can the order of the operations be clearly seen? ▪ Is the timing of the operations obvious? ▪ Is the design of the *works* and any Equipment shown, including acceptances? ▪ If the order and timing of the operations are required to be presented in a certain way by the Scope, does the programme reflect this? ▪ Does the programme look realistic and practicable? (Clause 31.3 provides that a reason for not accepting the programme is that it is not practicable or does not show the *Contractor's* plans realistically.)	
7	Order and timing of the work of the *Client* and Others as last agreed with them by the *Contractor* or, if not so agreed, as stated in the Scope ▪ Do the order and timing of the work of the *Client* and Others shown in the programme match the constraints in the Scope? ▪ If the directions in the Scope have been superseded by an agreement between the *Client*/Others, do the order and timing of the work of the *Client* and Others shown in the programme reflect that later agreement? (The *Client* and Others have no documentary relationship with the *Contractor*, and so the *Project Manager* should be able to put their hands on this superseding information: for example, an Early Warning Register which has resulted in a *Project Manager's* instruction; a *Project Manager's* instruction resulting from an ambiguity/inconsistency; any other *Project Manager's* instruction.)	

	Information to be included in the first programme	✓
8	Dates when the *Contractor* plans to meet each Condition stated for the Key Dates and to complete other work needed to allow the *Client* and Others to do their work ■ Does the programme show each Condition provided in the contract and the date when the Condition should be met? Do these dates occur on or before its correlating Key Date? ■ Does the programme show the dates for completion of the *Contractor's* other work that is a prerequisite for follow-on work to be done by the *Client* and Others? (This may link to point 7 above regarding the order and timing of the work of the *Client* and Others.)	
9	Provision for float ■ The *Contractor* is obliged to **show** provisions for float but is not obliged to **have** provision for float. The *Project Manager* may decide that the programme without float is not practicable or realistic.	
10	Provision for time risk allowances ■ The *Contractor* may feel that some operations are under risk of delay for whatever reason (delivery lead times, winter working, release of specialist Subcontractor from another project, etc.) and may choose to allow for this risk within their programme by building in extra time for those operations. ■ The *Contractor* is obliged to **show** provisions for time risk allowances but is not obliged to **have** such a provision. The *Project Manager* may decide that the programme without time risk allowances is not practicable or realistic.	
11	Provision for health and safety requirements ■ Does the programme show provision of health and safety requirements? (The *Contractor* is required to act in accordance with the health and safety requirements stated in the Scope (clause 27.4).) ■ Health and safety requirements will depend on the project. It is likely that any project that is complex or large enough to require a full programme will have some provisions for health and safety requirements, and the *Contractor* is required to show these on the programme submitted for acceptance.	
12	Provision for the procedures set out in the contract ■ Procedures may refer to those required by the clauses of the ECC (e.g. design acceptances, acceptance of Subcontractors, informing the *Supervisor* of tests/inspections, marking) and those required by the Scope (e.g. acceptance and procurement procedures (clauses C, D and E11.2(26) and F11.2(27)) and marking procedures (clause 71.1)).	
13	Dates when, in order to Provide the Works in accordance with their programme, the *Contractor* will need access to a part of the Site if later than its *access date* ■ Does the programme show when access to parts of the Site is required? ■ Does every operation show the access required? ■ The *Project Manager* needs to be aware of all access requirements to all parts of the Site. The *Client* is obliged to provide access on or before the *access date* stated in CD1 and the date for access that the *Contractor* shows on their programme that they submit for acceptance to the *Project Manager*. Although it is the *Client* who provides the access, it is the *Project Manager* who is managing the project and the Site and who will liaise with the *Client*.	
14	Dates when, in order to Provide the Works in accordance with their programme, the *Contractor* will need acceptances ■ The *Contractor* needs acceptances from the *Project Manager* and other people throughout the contract period. Does the programme show these acceptances? (For example, submission of design (clause 21.2), design of an item of Equipment (clause 23.1), and Subcontractors and their subcontract documents (clauses 26.2 and 26.3)).)	

	Information to be included in the first programme	✓
15	Dates when, in order to Provide the Works in accordance with their programme, the *Contractor* will need Plant and Materials and other things to be provided by the *Client* ■ Does the programme show when the *Contractor* will need things that are provided by Others? (For example, materials, facilities and samples for tests/inspections (clause 41.2), services and other things stated in the Scope (clause 25.2).)	
16	Dates when, in order to Provide the Works in accordance with their programme, the *Contractor* will need information from Others ■ Does the programme show when the *Contractor* will need information from Others? (For example, checks by the environmental agency and completion of the *works* of other contractors.)	
17	For each operation, a statement of how the *Contractor* plans to do the work, identifying the principal Equipment and other resources that they plan to use ■ Does the programme describe a means/method for the operation and the resources to be used?	
18	Other information that the Scope requires the *Contractor* to show on a programme submitted for acceptance ■ Does the programme show the information described in the Scope?	

NEC4: The Role of the *Project Manager*
ISBN 978-0-7277-6353-2

ICE Publishing: All rights reserved
http://dx.doi.org/10.1680/nectrpm.63532.135

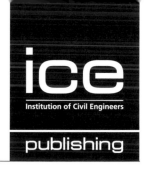

Appendix 15

Checklist for information to collect before undertaking the first assessment of the amount due

		Options	A	B	C	D	E	F
Information required for the first assessment of the amount due								
Activity Schedule	▪ Provided as part of the *Contractor's* tender		✓		✓			
Bill of Quantities	▪ Provided as part of the *Contractor's* tender			✓		✓		
Scope	▪ May describe what Completion of activities looks like; for comparison when determining the work is without Defects		✓		✓			
	▪ May describe activities to be undertaken separately from the Bill of Quantities; for comparison when determining the work is without Defects			✓		✓		
	▪ May describe further details of or requirements for the assessment of the amount due, e.g. documents, records and applications ▪ Acceptance and procurement procedures (not strictly part of Options A and B, as these Options do not have Disallowed Cost, but may be present for other reasons)		✓	✓	✓	✓	✓	✓
	▪ As a check against the Activity Schedule that all compensation events have been included in the Activity Schedule. (There should not be any at this early stage of an Option A contract, and any compensation events may not be at the end of their process; but there may be a few for Option C contracts.) The check is not required by the ECC but might be handy for the busy *Project Manager* to remind themselves of compensation events that have occurred but for which the process is not yet complete		✓		✓			
	▪ As a check against the Bill of Quantities that all compensation events have been included in the Bill of Quantities. (There should not be any at this early stage of an Option B contract, and any compensation events may not be at the end of their process; but there may be a few for Option D contracts.) This check is not required by the ECC but might be handy for the busy *Project Manager* to remind themselves of compensation events that have occurred but for which the process is not yet complete			✓		✓		
	▪ As a check for changes that have taken place						✓	✓

	Options	A	B	C	D	E	F
Information required for the first assessment of the amount due							
Accepted Programme	■ Has a first programme been submitted by the *Contractor* to the *Project Manager* for acceptance?	✓	✓	✓	✓	✓	
	■ To determine if a Key Date has been missed; and as information about upcoming activities that the *Contractor* will pay for over the next *assessment interval*			✓	✓	✓	✓
Any application for payment	■ Submitted by the *Contractor* to the *Project Manager* before the assessment date – clause 50.2	✓	✓	✓	✓	✓	✓
SCC	■ To be used as the basis for assessing the amount due – Defined Cost			✓	✓	✓	
CD1 by the *Client*	■ For information on secondary Options that could affect the first assessment of the amount due, e.g. Options X1 (not E or F), X3 (not C, D, E or F), X5, X6, X7, X14, X16 (not F), X17, X22 and Z	✓	✓	✓	✓	✓	✓
CD2 by the *Contractor*	■ Data for both schedules of cost components – to be used in Defined Cost ■ *fee percentage* ■ *working areas*			✓	✓	✓	
	■ The description of work that the *Contractor* will do themselves and the *price* per activity described ■ *working areas*						✓
Details of the banking arrangements	■ For the Project Bank Account submitted by the *Contractor* to the *Project Manager* under secondary Option Y(UK)1	✓	✓	✓	✓	✓	✓
Records of amounts that the *Contractor* has paid to their Subcontractors	■ These records must be of sufficient detail for the *Project Manager* to be able to isolate the payments described in clause 11.2(24) (e.g. retention and payment to Others). The records may have been submitted by the *Contractor* under clause 50.2			✓	✓	✓	✓

NEC4: The Role of the *Project Manager*
ISBN 978-0-7277-6353-2

ICE Publishing: All rights reserved
http://dx.doi.org/10.1680/nectrpm.63532.137

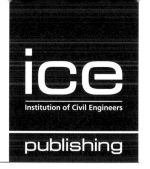

Checklist for the first assessment of the amount due

Options	A	B	C	D	E	F
First assessment of the amount due – all main Options						
1 Price for Work Done to Date	✓	✓	✓	✓	✓	✓
1a Are any of the activities on the Activity Schedule complete? ▨ Look at the descriptions of the activities and groups of activities on the Activity Schedule (remember that if activities are in a group, all activities within the group must be complete before the group can be assessed as complete) ▨ Compare the Activity Schedule with the Accepted Programme in case of anomalies ▨ Check the Defect notification log to confirm that there are no Defects that would either delay or be covered by immediately following work	✓					
Using each item in the Bill of Quantities ▨ Assess the quantity that has been completed for each item ▨ Work out how complete the *works* are in percentage terms to allocate the appropriate portion of any lump sums ▨ Check the Defect notification log to confirm that there are no Defects that would either delay or be covered by immediately following work		✓				
1b Check the Activity Schedule for the Prices to be used in calculating the Price for Work Done to Date	✓					
1c Has anything happened to change the Prices on the Activity Schedule? In other words, is the Activity Schedule up to date and accurate? ▨ Compensation events (check the compensation event log) ▨ Acceleration ▨ Acceptance of Defects ▨ Change to a planned method of working	✓					
What events since the *starting date* have changed the Bill of Quantities? ▨ Compensation events (check the compensation event log) ▨ Acceleration ▨ Acceptance of Defects ▨ Change to a planned method of working		✓				
1d Are there other activities that may affect the assessment of the amount due? For example ▨ Equipment, Plant and Materials that have been marked as required by clause 71.1 and are identified for payment ▨ Tests/inspections that have been successful and a conditional payment has become due	✓	✓				

	Options	A	B	C	D	E	F
	First assessment of the amount due – all main Options						
1e	Mark three points on the Accepted Programme (and, for Option C, the Activity Schedule) ■ The point reached by the assessment date ■ The point that will be reached by the next assessment date (as a forecast) ■ The forecast implementation date of any compensation events for which the compensation event procedure has already started and that are likely to reach conclusion before the next assessment date			✓	✓	✓	✓
1f	Check the Activity Schedule for the Prices shown to get an indication of intended expenditure. The Price for Work Done to Date should not be all that different from the Prices at this early stage of the project			✓			
	Using each item in the Bill of Quantities ■ Count the quantity that has been completed for each item ■ Work out how complete the *works* are in percentage terms to allocate the appropriate portion of any lump sums to get an idea of intended expenditure at this point in the project				✓		
1g	Gather any documentation on any compensation events to see if they will affect this first assessment			✓	✓	✓	✓
1h	Set aside time to go through the *Contractor's* application for payment, matching it with the documents submitted as evidence			✓	✓	✓	✓
1i	Set up a tick document in a spreadsheet or other software with the list of items in clause 11.2(24)[a] and 11.2(26)[a] and the list of costs in the SCC so that each one is considered and nothing is erroneously omitted [a] Clauses 11.2(25) and 11.2(27) for Option F			✓	✓	✓	✓
1j	Set up the project folders so that scanned evidence can be stored electronically by the assessment date and easily accessed for audits or other checks			✓	✓	✓	✓
2	Plus other amounts to be paid to the *Contractor* less amounts to be paid by or retained from the *Contractor*	✓	✓	✓	✓	✓	✓
	■ Option X1, 'Price adjustment for inflation': unlikely to be required in the first assessment of the amount due	✓	✓	✓	✓		
	■ Option X2, 'Changes in the law': unlikely to be required in the first assessment of the amount due	✓	✓	✓	✓	✓	✓
	■ Option X3, 'Multiple currencies': a simple conversion is required	✓	✓				
	■ Option X5, 'Sectional Completion': unlikely to be required in the first assessment of the amount due	✓	✓	✓	✓	✓	✓
	■ Option X6, 'Bonus for early Completion': unlikely to be required in the first assessment of the amount due	✓	✓	✓	✓	✓	✓
	■ Option X7, 'Delay damages': unlikely to be required in the first assessment of the amount due	✓	✓	✓	✓	✓	✓
	■ Option X14, 'Advanced payment to the *Contractor*': the appropriate amount may already have been paid under clause X14.2, or it may be part of the first assessment of the amount due (e.g. if the date of the first assessment is less than 4 weeks after the Contract Date or if the *Client* has already received the required advanced payment bond)	✓	✓	✓	✓	✓	✓
	■ Option X16, 'Retention': see item 2a below	✓	✓	✓	✓	✓	

		Options	A	B	C	D	E	F
	First assessment of the amount due – all main Options							
	▪ Option X17, 'Low performance damages': unlikely to be required in the first assessment of the amount due		✓	✓	✓	✓	✓	✓
	▪ Option X20, 'Key Performance Indicators': unlikely to be required in the first assessment of the amount due		✓	✓	✓	✓	✓	✓
	▪ Option Z, '*Additional conditions of contract*': check if any Option Z clauses affect the amount due		✓	✓	✓	✓	✓	✓
2a	Option X16, 'Retention' ▪ If the *retention free amount* provided in CD1 by the *Client* is greater than zero, check whether the Price for Work Done to Date assessed has reached the *retention free amount*. If it has not, there is no further action required by the *Project Manager* ▪ If the Price for Work Done to Date assessed has reached the *retention free amount*, then the *Project Manager* must assess a retention from the amount due equal to the mathematical product of the *retention percentage* multiplied by the portion of the Price for Work Done to Date that is greater than the *retention free amount*		✓	✓	✓	✓	✓	

NEC4: The Role of the *Project Manager*
ISBN 978-0-7277-6353-2

Appendix 17
Checklist for early warnings

	Items to check on receiving or providing a notification for an early warning	✓
1	General information: ■ number ■ date ■ description of the matter	
2	Effect of the matter – will it ■ increase the total of the Prices? ■ delay Completion? ■ delay meeting a Key Date? ■ impair the performance on the *works* in use? ■ increase the *Contractor's* total cost? ■ affect any other matter?	
3	Description of how the matter arose and its effect on the project and the people	
4	Is the matter one of the *Client's* liabilities (clauses 60.1(14) and 80.1 and CD1)?	
5	Is the matter likely to become a compensation event?	
6	Ideas for mitigation/avoidance and solutions	
7	In addressing the early warning matter, is a special early warning meeting required or can the matter be addressed at the project's regular early warning meeting?	

NEC4: The Role of the _Project Manager_
ISBN 978-0-7277-6353-2

ICE Publishing: All rights reserved
http://dx.doi.org/10.1680/nectrpm.63532.143

Appendix 18
Agenda for the monthly meeting between the _Project Manager_ and the _Supervisor_

1. Quality management system (clause 40)
    - Has the _Contractor's_ quality plan changed?
    - Has the _Contractor_ failed to comply with the quality plan?

2. Tests/inspections before delivery (clause 42.1)
    - Plant and Materials which have passed tests and inspections and can be delivered to the Working Areas

3. Marking of Equipment, Plant and Materials outside the Working Areas (clause 71.1)
    - Equipment, Plant and Materials which are outside the Working Areas and which the _Supervisor_ has marked for payment in accordance with the Scope

4. Payment – discussion about whether activities are complete
    - Activities and work which is complete, i.e. without Defects that would either delay or be covered by immediately following work
    - Defects to be corrected after Completion
    - Defects caused by the _Contractor_ not complying with a constraint on how it is to Provide the Works stated in the Scope

5. Early warnings and compensation events (clauses 15 and 60.1)
    - Discussions about early warnings and other risks
    - Any _Supervisor_ actions that may be compensation events
        - _Supervisor_ does not reply in time
        - _Supervisor_ has changed a decision
        - A search was instructed and no Defect found
        - Unnecessary delay on a test or inspection
    - Any changes to the Accepted Programme as a result of a compensation event that may affect the _Supervisor's_ duties or schedules

6. Site information and records
    - Discussions about any Site information recorded during the previous month

NEC4: The Role of the *Project Manager*
ISBN 978-0-7277-6353-2

ICE Publishing: All rights reserved
http://dx.doi.org/10.1680/nectrpm.63532.145

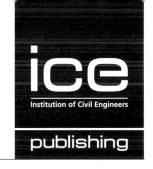

Appendix 19
Checklist for information to be included in subsequent programmes

	List of dates/points in time/information that the *Contractor* must include in subsequent programmes submitted for acceptance	✓
1	*Access dates* ■ Have any parts of the *works* yet to start? ■ Are the *access dates* included in the programme still realistic and achievable?	
2	Key Dates ■ Are any Key Dates still programmed and achievable? ■ Are any *Project Manager's* instructions changing the Key Dates reflected on the programme?	
3	Completion Date ■ Is the *completion date* for the whole of the *works* which is shown on the programme the latest agreed date, taking into account compensation events, notified Defects and acceleration? For example, the correction of a notified Defect that would prevent Completion of the *works* could be included on the programme to show that it will be corrected before the *completion date*.	
4	Date of planned Completion ■ Is the date for planned Completion still on or before the Completion Date?	
5	Order and timing of the operations that the *Contractor* plans to do in order to Provide the Works ■ Are the operations shown on the programme still representative of the *works*, with the actual progress achieved shown for each operation as well as the effect upon the timing of the remaining work? ■ Do the order and timing of the operations show the effects of implemented compensation events? ■ Do the order and timing of the work show how the *Contractor* plans to deal with any delays and how they plan to correct notified Defects? ■ Do the operations show any changes that the *Contractor* plans to make to the Accepted Programme? ■ Are any *Project Manager's* instructions reflected in the operations? ■ Have any accepted changes to the Working Areas been reflected in the operations?	
6	Order and timing of the work of the *Client* and Others as last agreed with them by the *Contractor* or, if not so agreed, as stated in the Scope ■ Do the order and timing of the work of the *Client* and Others shown in the programme match the constraints in the Scope? ■ If the directions in the Scope have been superseded by an agreement between the *Client*/Others, do the order and timing of the work of the *Client* and Others shown in the programme reflect that later agreement? (The *Client* and Others have no documentary relationship with the *Contractor*, and so the *Project Manager* should be able to put their hands on this superseding information: for example, an Early Warning Register that has resulted in a *Project Manager's* instruction; a *Project Manager's* instruction resulting from an ambiguity/inconsistency or changing the Scope; any other *Project Manager's* instruction.)	

	List of dates/points in time/information that the *Contractor* must include in subsequent programmes submitted for acceptance	✓
7	Dates when the *Contractor* plans to meet each Condition stated for the Key Dates and to complete other work needed to allow the *Client* and Others to do their work ■ Does the programme show each Condition provided in the contract and the date when the Condition should be met? Does each of these dates occur on or before its correlating Key Date? ■ Does the programme show each Condition that has already been met and the date on which it was met? ■ Does the programme show the dates for completion of the *Contractor's* other work that is a prerequisite for follow-on work to be done by the *Client* and Others? (This may link to point 6 above regarding the order and timing of the work of the *Client* and Others.) ■ Does the programme show dates when activities or other pieces of work were complete?	
8	Provision for float ■ Has the *Contractor* used any of their float shown on the programme? ■ How does this affect the other programmed activities? ■ Does the *Contractor* have other float that may be used in case of delay and is it realistic given the progress to date?	
9	Provision for time risk allowances ■ Has the *Contractor* used some of their time risk allowances? ■ Are other time risk allowances included in the programme and are they still realistic given progress to date?	
10	Provision for health and safety requirements ■ Does the programme show provision of health and safety requirements, and are they realistic given changes to the *works*, e.g. compensation events?	
11	Provision for the procedures set out in the contract ■ Are the procedures still taken into account given any changes to the *works*?	
12	Dates when, in order to Provide the Works in accordance with their programme, the *Contractor* will need access to a part of the Site if later than its *access date* ■ Does the programme show when access to parts of the Site is required? ■ Does every operation show the access required?	
13	Dates when, in order to Provide the Works in accordance with their programme, the *Contractor* will need acceptances ■ Does the programme show the required acceptances?	
14	Dates when, in order to Provide the Works in accordance with their programme, the *Contractor* will need Plant and Materials and other things to be provided by the *Client* ■ Does the programme show when the *Contractor* will need things that are provided by Others? ■ Are the changes to the *works* and the programme reflected in the programme?	
15	Dates when, in order to Provide the Works in accordance with their programme, the *Contractor* will need information from Others ■ Does the programme show when the *Contractor* will need information from Others and are any changes reflected in the programme?	

	List of dates/points in time/information that the *Contractor* must include in subsequent programmes submitted for acceptance	✓
16	For each operation, a statement of how the *Contractor* plans to do the work, identifying the principal Equipment and other resources that they plan to use ■ Does the programme describe a means/method for the operation and the resources to be used? ■ Are changes to the *works* (e.g. compensation events) accurately reflected? ■ Are the operation statements still realistic given progress to date?	
17	Other information that the Scope requires the *Contractor* to show on a programme submitted for acceptance ■ Does the programme show the information described in the Scope, including any changes resulting from compensation events?	

NEC4: The Role of the *Project Manager*
ISBN 978-0-7277-6353-2

ICE Publishing: All rights reserved
http://dx.doi.org/10.1680/nectrpm.63532.149

Checklist for information to be included in the *Contractor's* submitted programme

	Clause	*Project Manager's* checks on the *Contractor's* submitted programme
1	11.2(2) 'Completion is when the *Contractor* has done all the work which the Scope states is to be done by the Completion Date'	■ Check the Scope: – Does it state what work is to be done by the Completion Date? ■ Check the programme: – Does it show all this work on the programme? – Is the date of Completion marked as being after all this work has been done?
2	20.1 'Provides the Works in accordance with the Scope'	Does the programme show the *works* identified in the Scope?
3	21.1 'The *Contractor* designs the parts of the *works* which the Scope states the *Contactor* is to design'	Does the programme show the timing for submission of the design for acceptance?
4	27.4 'The *Contractor* acts in accordance with the health and safety requirements stated in the Scope'	Does the programme make provision for any health and safety requirements that apply?
5	41.2 'The *Contractor* and the *Client* provide materials, facilities and samples for tests and inspections as stated in the Scope'	If the Scope provides information about supplies for tests/inspections, does the programme show the test/inspection so that the *Client* can provide the supplies at the right time?
6	42.1 'Plant and Materials which the Scope states are to be tested or inspected before delivery'	Does the programme appear to recognise that the delivery of the relevant Plant and Materials and their use on Site or in the Working Areas takes into account time for the *Supervisor's* tests and inspections?
Core clauses		
7	11.2(2) 'Completion is when the *Contractor* has done all the work necessary for the *Client* to use the *works* and for Others to do their work'	Is the date of Completion provided on the programme marked as being after all the work has been done that is necessary for the *Client* to use the *works* and for Others to do their work?

	Clause	*Project Manager's* checks on the *Contractor's* submitted programme
8	Core clause section 3 'Time'	Most of the clauses relating to the programme are provided in this part of the ECC.
9	23.1 'The *Contractor* submits particulars of the design of an item of Equipment'	Although this submission is only required if the *Project Manager* instructs it, the *Project Manager* may want to ascertain what time has been provided for carrying out this task by the *Contractor*.
10	27.1 'The *Contractor* obtains approval of its design from Others where necessary'	Does the programme show the timing of approval from Others? (Note: To be thorough, the *Project Manager* may want to check whether Others need to check the *Contractor's* design so that they can ensure that the programme has left enough time to undertake this requirement. It may be that the Scope contains appropriate information.)
11	41.3 'The *Contractor* informs the *Supervisor* in time for a test or an inspection to be arranged and done before doing work which would obstruct the test or inspection'	Does the programme show that the *Contractor* has left sufficient time to notify the required people and to undertake the test without causing delay?
12	41.5 'The *Supervisor* does tests and inspections without causing unnecessary delay'	Does the programme provide time for the *Supervisor* to make the notifications required by clause 41.3 and to do the test/inspection promptly for each test/inspection required?
13	44.2 'The *Contractor* corrects a notified Defect'	Although the programme cannot take into account time for specific Defects, does it appear to include some float for the correction of any Defects that would prevent Completion of the *works*?
14	50.3 'The Price for Work Done to Date'	Does the programme show that the operations are sensibly placed within time and are easy to identify so that the *Contractor's* cash flow is assured throughout the project duration?
15	60.1 Compensation events: (2), (3), (4), (5), (11), (15), (16) and (19)	The *Project Manager* cannot check for these compensation events on the programme, but all of them rely on the Accepted Programme, even if just as a check for a delay (e.g. (11)) or against points of time (e.g. (15)).
16	62.1 'After discussing with the *Contractor* different ways of dealing with the compensation event'	The *Project Manager* will use the Accepted Programme when reviewing any compensation event to explore how the Accepted Programme will be affected by different scenarios.
17	63.5 'A delay to the Completion Date is assessed as ... planned Completion is later than planned Completion as shown on the Accepted Programme'	If planned Completion is not shown on the Accepted Programme, then the *Contractor* will have difficulty in presenting their quotations for compensation events.

	Clause	*Project Manager's* checks on the *Contractor's* submitted programme
Optional clauses		
18	Option A clauses Use of the Activity Schedule	Does the programme correlate with the Activity Schedule?
19	A31.4 and A55.3 'The *Contractor* … shows how each activity on the Activity Schedule relates to the operations on each programme'	Is each activity represented in the programme?
20	C, D, E, F20.4 'The *Contractor* prepares forecasts of the total Defined Cost for the whole of the *works*'	Do the forecasts correlate with the Accepted Programme?
21	Option X5 *section* of the *works*	Check that the programme reflects the *sections* of the *works* provided for in the Contract Data.
22	Option X6 Completion and the Completion Date	Check what the *Contractor* is planning with regards to Completion and, with this in mind, check the practicability of the programme.
23	Option X14 Contract Date	Check that the programme contains enough information to support the dates required by Option X14.
CD1 by the *Client*		
24	11.2(11) The *key dates* and *conditions* are those stated in the Contract Data (unless later changed)	Provided in the programme?
25	The *starting date* The *access dates* The *completion date* for the whole of the *works*	Provided in the programme?
26	The *Contractor* submits revised programmes at intervals	CD1 identifies the frequency (usually monthly).
27	The period after the Contract Date within which the Contractor is to submit a first programme for acceptance	Where a first programme is not provided with the tender in CD2 by the *Contractor* and instead is provided after the contract has been awarded.
28	The *key dates* and *conditions* to be met	Does the programme show the Key Dates and Conditions? (See clause 11.2(11).)
29	Option X5 The *completion date* for each *section* of the *works*	Provided in the programme?
CD2 by the *Contractor*		
30	The programme identified in the Contract Data	Where a first programme is provided with the tender.

NEC4: The Role of the *Project Manager*
ISBN 978-0-7277-6353-2

ICE Publishing: All rights reserved
http://dx.doi.org/10.1680/nectrpm.63532.153

Appendix 21
Checklist for information to collect before undertaking assessments of the amount due

Options		A	B	C	D	E	F
Information required for second and later assessments of the amount due							
Up-to-date Activity Schedule	Taking into account ■ CD2 by the *Contractor* – tendered total of the Prices ■ The compensation event log and the resulting changes to the Prices and the Activity Schedule ■ The latest Accepted Programme and its impact on the Activity Schedule (e.g. acceleration)	✓		✓			
Up-to-date Bill of Quantities	Taking into account ■ CD2 by the *Contractor* – tendered total of the Prices ■ The compensation event log and the resulting changes to the Prices and the Bill of Quantities ■ The latest Accepted Programme and its impact on the Bill of Quantities (e.g. acceleration)		✓		✓		
Previous Price for Work Done to Date	Taking into account ■ the *Project Manager's* previous payment certificates, which will build on each other to provide the baseline for the subsequent assessment	✓	✓	✓	✓	✓	✓
Scope	■ As for the first assessment	✓	✓	✓	✓	✓	✓
Compensation event log	■ As for the first assessment	✓	✓	✓	✓	✓	✓
The Accepted Programme	■ As for the first assessment	✓	✓	✓	✓	✓	✓
Any application for payment	■ As for the first assessment	✓	✓	✓	✓	✓	✓
SSC	■ As for the first assessment			✓	✓	✓	
CD1 by the *Client*	■ As for the first assessment, except that some of the Options are more likely to apply in later assessments (e.g. Completion of *sections* of the *works* with associated Options X6 and X7, Retention in Option X16 and Key Performance Indicators in Option X20).	✓	✓	✓	✓	✓	✓
CD2 by the *Contractor*	■ As for the first assessment			✓	✓	✓	
Records of amounts that the *Contractor* has paid to their Subcontractors	■ As for the first assessment			✓	✓	✓	✓

NEC4: The Role of the *Project Manager*
ISBN 978-0-7277-6353-2

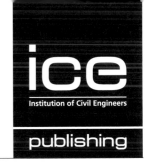

Checklist for assessments that take place outwith core clause section 5

Assessments that take place outwith core clause section 5
Plus other amounts to be paid to the *Contractor* **less** amounts to be paid by or retained from the *Contractor*.
Examples

Clause 25.2

■ The *Project Manager* **assesses any cost incurred by the *Client*** as a result of the *Contractor* not providing the service and other things that they are to provide.

In preparation for this assessment, the *Project Manager* could give the *Contractor* notification that they have not provided something they should have. Where the provision is time based (i.e. the *Contractor* provided the things but not within the timescales required), the *Project Manager* may wish to remind the *Contractor*, at the weekly progress meeting prior to the time when the things are due, that the things are required and that there are consequences to late provision.

Clause 25.3

■ The *Project Manager* is allowed to decide that the *Contractor's* work has not met the Conditions stated for the Key Date by the date stated and that, as a result, the *Client* has incurred an additional cost on the same contract.

The *Client* must choose whether to carry out the work themselves or to pay Others to do it (e.g. another contractor who is already on Site or who is brought in especially to meet the Condition). The *Project Manager* has 4 weeks after the Condition for the Key Date is met (by the *Client* or Others) in which to **assess the additional cost to the *Client***. If the *Contractor* carries out the work and later meets the Condition (generally this would be the best solution for both Parties) then there are no consequences other than those provided in the secondary Options.

Clause 41.6

■ The *Project Manager* must **assess the cost incurred by the *Client*** if a test/inspection has to be repeated after a Defect is found.

The *Project Manager* may need to liaise with the *Supervisor* to understand the costs to the *Client* in repeating the test/inspection.

Clause 46.1

■ Where the *Contractor* is given access to correct a notified Defect after take over but does not correct it within their *defect correction period*, the *Project Manager* must **assess the cost incurred by the *Client*** of having a Defect correction carried out by other people.

There are obvious differences between the *Contractor* (1) not attempting to correct the Defect, (2) attempting a correction but being unable to do so, or (3) correcting the Defect but taking longer than the relevant *defect correction period*. In general, the best solution is for the *Contractor* to correct the Defect, even if it takes a bit longer than the *defect correction period*, and the *Project Manager* after discussion with the *Client* may want to consider some leniency.

Assessments that take place outwith core clause section 5
Clause 46.2
■ Where the *Contractor* is not given access to correct a notified Defect after take over, the *Project Manager* must **assess the cost** that would have been incurred by the *Contractor* of correcting that Defect. The *Project Manager* may need to liaise with the *Supervisor* to understand fully the costs to the *Contractor* in correcting the Defect.
Clause 93.1
■ The *Project Manager* must **assess the amount due** on termination in the manner described in clause 93.1. The *Project Manager's* assessment will be very similar to the regular assessments that take place during the contract period.

NEC4: The Role of the *Project Manager*
ISBN 978-0-7277-6353-2

ICE Publishing: All rights reserved
http://dx.doi.org/10.1680/nectrpm.63532.157

Checklist for information to collect before undertaking the assessment of the amount due after Completion

	Options	A	B	C	D	E	F
Information required for the next assessment of the amount due after Completion of the whole of the *works*: in addition to previous assessments							
Options that are affected by Completion	For example, Options X6, X7, X16, X17, X20 and X22	✓	✓	✓	✓	✓	✓
Accepting a Defect: clause 45.2	Instructions changing the Scope, the Prices and the Completion Date	✓	✓	✓	✓	✓	✓
Uncorrected Defects: clause 46.1	Assesses the cost of having the Defect corrected by other people	✓	✓	✓	✓	✓	✓
Uncorrected Defects: clause 46.2	Assesses the cost of correcting the Defects	✓	✓	✓	✓	✓	✓

NEC4: The Role of the *Project Manager*
ISBN 978-0-7277-6353-2

Appendix 24
Checklist for assessment of the amount due after Completion of the whole of the *works*

The next assessment of the amount due after Completion of the whole of the *works* – all main Options		✓
1	**Price for Work Done to Date** ■ Final activities on the Activity Schedule – for example, the delivery of O&M manuals or as-built drawings ■ Perhaps an activity tied to the passing of final tests or commissioning	
2	**Plus** other amounts to be paid to the *Contractor* **less** amounts to be paid by or retained from the *Contractor*	
2a	Option X7 Delay damages Facts: ■ The date of Completion certified by the *Project Manager* was 6 December 20XX ■ The date of take over certified by the *Project Manager* was 9 December 20XX CD1 by the *Client*: ■ The *completion date* for the whole of the *works* is 2 December 20XX ■ Delay damages for Completion of the whole of the *works* are £250 per day Therefore: ■ Delay damages are due at £250 per day from 2 to 6 December (4 days) = £1000	
2b	Option X16 Retention Assumptions (for the purpose of this example): CD1 by the *Client*: ■ The *completion date* for the whole of the *works* is 2 December 20XX Secondary Option X16 Retention ■ The retention-free amount is £0 ■ The retention percentage is 5% CD2 by the *Contractor*: ■ The tendered total of the Prices is £173 445 Therefore: ■ The Price for Work Done to Date in the next assessment following Completion of the whole of the *works* is £194 000. The retention amount is halved in this assessment e.g. Retention on the PWDD is 5% of £194 000 = £9700/2 = £4850. The retention remains at this amount until the date when the Defects Certificate is due to be issued. No amount is retained in the assessments made after the Defects Certificate is due to be issued	

	The next assessment of the amount due after Completion of the whole of the *works* – all main Options	✓
2c	Option X20, 'Key Performance Indicators'	
	Main Options C and D	
	Contractor's share ■ Preliminary assessment of the *Contractor's share*	
2d	Option X22, 'Early *Contractor* involvement (used only with Options C and E)'	
	Main Options C and E	
	■ Preliminary assessment of the *budget incentive* at Completion of the whole of the *works* included this in the amount due following Completion of the whole of the *works*	

NEC4: The Role of the *Project Manager*
ISBN 978-0-7277-6353-2

ICE Publishing: All rights reserved
http://dx.doi.org/10.1680/nectrpm.63532.161

Appendix 25
Checklists for Completion and take over

(a) Early take over of a *section* of the *works* (Option X5, 'Sectional completion')		✓
1	Is the *Client* willing to take over a *section* of the *works* before the Completion Date (CD1)?	
2	Does the *Client* want to take over this *section* of the *works* before the Completion Date?	
3	Has the *Client* already started using this *section* of the *works* even though Completion has not been certified?	
4	Does the Scope include a reason for the *Client* to take over this section of the *works* before Completion?	
5	Is the reason for the *Client* taking over this *section* of the *works* one of the reasons stated in the Scope?	
6	Therefore, has early take over of that *section* of the *works* occurred?	
7	What is the date of take over for that *section* of the *works*?	
8	Certify the date upon which the *Client* takes over the *section* of the *works* and its extent within 1 week of the date on which take over of that *section* of the *works* is deemed to have taken place	
9	Include the date on the take over certificate, and issue the take over certificate for that *section* of the *works* to the *Client* and the *Contractor* within 1 week of the date on which take over for that section of the *works* occurred	
10	If Option X6 applies: the *Contractor's* bonus is calculated from the date of take over of the *section* of the *works* (the date is certified by the *Project Manager* on the take over certificate) to the Completion Date for that *section* of the *works* (stated on the Accepted Programme)	

(b) Completion of a *section* of the *works* (Option X5, 'Sectional completion')		✓
1	Has the *Contractor* done all the work necessary for the *Client* to use this *section* of the *works* and for Others to do their work?	
2	Does the Scope provide enough information about what work is to be done before the Completion Date of this *section* of the *works*?	
3	Does the *Project Manager* have access to the list of notified Defects?	
4	Will any of the uncorrected notified Defects prevent the *Client* from using this section of the *works*?	
5	Will any of the uncorrected notified Defects prevent Others from doing their work?	
6	What is the date on which all the conditions were met and thus what is the date on which Completion of this *section* of the *works* took place?	

(b) Completion of a *section* of the *works* (Option X5, 'Sectional completion')	✓
7	Include the date on the certificate of Completion for the *section* of the *works* and issue the Completion certificate to the *Client* and the *Contractor* within 1 week of the date on which the *Project Manager* decided Completion of this *section* of the *works* has occurred.	
8	If Option X6 Bonus for early Completion applies: If Completion of a *section* of the *works* takes place before take over of that *section*, the *Contractor's* bonus is calculated from the date of Completion of the *section* (the date is certified by the *Project Manager* on the Completion certificate) to the Completion Date of the *section* (stated on the Accepted Programme).	

(c) Take over of a *section* of the *works* where Completion has already been reached and early take over has not taken place (Option X5, 'Sectional completion')	✓
1	What is the date of Completion for that *section* of the *works*?	
2	Take over of a section of the *works* occurs not later than 2 weeks after Completion of that *section*	
3	Within 1 week of the date of take over of a *section*: ■ certify the date upon which the *Client* takes over that *section* of the *works* and its extent ■ issue the take over certificate for that *section* of the *works* to the *Client* and the *Contractor*.	

(d) Early take over of the whole of the *works* (clause 35)	✓
1	Is the *Client* willing to take over the whole of the *works* before the Completion Date?	
2	Does the *Client* want to take over the whole of the *works* before the Completion Date?	
3	Has the *Client* already started using parts of the *works* even though Completion of the whole of the *works* has not been certified?	
4	Does the Scope include a reason for the *Client* to take over the whole of the *works* before Completion?	
5	Is the reason for the *Client* taking over the whole of the *works* one of the reasons stated in the Scope?	
6	Has early take over of the whole of the *works* occurred?	
7	What is the date of take over of the *works*?	
8	Certify the date upon which the *Client* takes over the *works* and its extent within 1 week of the date on which take over is deemed to have taken place	
9	Include the date on the take over certificate, and issue the take over certificate to the *Client* and the *Contractor* within 1 week of the date on which take over occurred	
10	If Option X6 ('Bonus for early Completion') applies: The *Contractor's* bonus is calculated from the date of take over of the *works* (the date is certified by the *Project Manager* on the take over certificate) to the Completion Date for the *works* (stated on the Accepted Programme)	
11	If Option X16 ('Retention') applies: ■ from the take over date the *Client* no longer retains the *retention percentage* in each Price for Work Done to Date ■ in the next assessment after the *Client* has taken over the whole of the *works*, the amount of retention is halved	

(e) Completion of the whole of the *works*		✓
1	Has the *Contractor* done all the work necessary for the *Client* to use the *works* and for Others to do their work?	
2	Does the Scope provide enough information about what work is to be done before the Completion Date?	
3	Does the *Project Manager* have access to the list of notified Defects?	
4	Will any of the uncorrected notified Defects prevent the *Client* from using the *works*?	
5	Will any of the uncorrected notified Defects prevent Others from doing their work?	
6	What is the date on which all the conditions were met and thus what is the date on which Completion took place?	
7	Include the date on the certificate of Completion, and issue the Completion certificate to the *Client* and the *Contractor* within 1 week of the date on which the *Project Manager* decided Completion has occurred	
8	Undertake the assessment of the amount due at the next assessment date after Completion of the whole of the *works*	
9	If Option X6 ('Bonus for early Completion') applies: if Completion of the *works* takes place before take over, the *Contractor's* bonus is calculated from the date of Completion of the *works* (the date is certified by the *Project Manager* on the Completion certificate) to the Completion Date of the *works* (stated on the Accepted Programme)	
10	Calculate the Price Adjustment Factor at the Completion Date for the whole of the *works* (Option X1)	
11	If Option X16 ('Retention') applies and Completion of the whole of the *works* takes place before take over of the whole of the *works*: ■ Completion marks the end of the period during which the *Client* retains the *retention percentage* in each Price for Work Done to Date ■ in the next assessment after Completion of the whole of the *works*, the amount of retention is halved	

(f) Take over of the whole of the *works* where Completion of the whole of the *works* has already been reached and early take over has not taken place		✓
1	What is the date of Completion for the whole of the *works*?	
2	Take over of the *works* occurs not later than 2 weeks after Completion of the whole of the *works*	
3	Within 1 week of the date of take over: ■ certify the date upon which the *Client* takes over the *works* and its extent ■ issue the take over certificate to the *Client* and the *Contractor*	

NEC4: The Role of the *Project Manager*
ISBN 978-0-7277-6353-2

ICE Publishing: All rights reserved
http://dx.doi.org/10.1680/nectrpm.63532.165

Checklist for the *defects date* and the issue of the Defects Certificate

Things to do at the *defects date* and issue of the Defects Certificate	✓
1	The *Supervisor* issues their Defects Certificate to the *Project Manager* and *Contractor* (clause 13.6)	
2	A compensation event is not notified after the *defects date* (clause 61.7)	
3	The *Project Manager* makes an assessment of the amount due no later than 4 weeks after the issue of the Defects Certificate (clause 53.1)	
4	Final release of any remaining retention held. No retention is retained in the assessment after the Defects Certificate is due to be issued (clause X16.2)	
5	The *Project Manager's* assessment of the final amount due takes place no later than 4 weeks after the issue of the Defects Certificate, and takes into account any Defects listed on the Defects Certificate as not having been corrected	
6	If a Defect shows low performance, then the *Contractor* pays the low-performance damages stated in the Contract Data (clause X17.1)	
7	The *Contractor's* liability for Defects due to its design which are not listed on the Defects Certificate is limited to the amount stated in the Contract Data (clause X18.4)	
8	The *Project Manager* will not receive any further reporting from the *Contractor* after the issue of the Defects Certificate (clause X20.2 – not used with Option X12)	

NEC4: The Role of the *Project Manager*
ISBN 978-0-7277-6353-2

ICE Publishing: All rights reserved
http://dx.doi.org/10.1680/nectrpm.63532.167

Appendix 27

Information that the *Project Manager* needs to consider for the final assessment

	Options	A	B	C	D	E	F
Assessment of the final amount due no later than 4 weeks after the *Supervisor* issues the Defects Certificate							
Defects Certificate	Check whether the Defects Certificate issued by the *Supervisor* lists any Defects that were notified before the *defects date* but not corrected	✓	✓	✓	✓	✓	✓
Accepting a Defect: clause 45.2	Instructions changing the Scope, the Prices, Key Dates and the Completion Date	✓	✓	✓	✓	✓	✓
Uncorrected Defects: clause 46	Adjustment to the amount due for uncorrected Defects	✓	✓	✓	✓	✓	✓
Option X1, 'Price adjustment for inflation'	An amount for price adjustment is added to the total of the Prices	✓	✓	✓	✓		
Option X16, 'Retention'	Release of the second half of the retention if secondary Option X16 was part of the contract	✓	✓	✓	✓	✓	
Option X17, 'Low performance damages'	If a Defect in the Defects Certificate shows low performance	✓	✓	✓	✓	✓	✓
Option X20, 'Key Performance Indictors (not used with Option X12)'	Performance against Key Performance Indicators	✓	✓	✓	✓	✓	✓
Main Options C and D	Final assessment of the *Contractor's* share			✓	✓		
Main Options C, D, E and F	Insurance – adjustment for costs paid by insurers to the *Contractor*			✓	✓	✓	✓

NEC4: The Role of the *Project Manager*
ISBN 978-0-7277-6353-2

ICE Publishing: All rights reserved
http://dx.doi.org/10.1680/nectrpm.63532.169

Appendix 28
Checklist for the final assessment of the amount due

Final assessment of the amount due – all main Options	✓
1 Price for Work Done to Date	
1a It is unlikely that there will be anything included in the assessment of the amount due at this stage of the project, particularly since the cost of Defects that are corrected after Completion is specifically excluded from the Price for Work Done to Date	
2 Plus other amounts to be paid to the *Contractor* less amounts to be paid by or retained from the *Contractor* This could include assessments of ▪ acceptance of Defects (clause 45) ▪ uncorrected Defects (clause 46)	
2a Option X16, 'Retention (not used with Option F)' Facts: CD1 provided by the *Client*: ▪ The retention-free amount is £0 ▪ The retention percentage is 5% Therefore: ▪ In the final assessment of the amount due, the *Project Manager* will include for payment the amount of retention that has been held up to that point ▪ If the Price for Work Done to Date at Completion was £192 665, an amount of £9633 will have been retained up to that point and an amount of £4816 released in the next assessment that took place after Completion, leaving another £4816 to be released at the assessment of the final amount due	

2b Option X17, 'Low performance damages'
 	
Facts:
 	
▪ The Defects Certificate lists a Defect that shows low performance with respect to a performance level stated in the Contract Data
 	
CD1 by the *Client*:
 	
▪ The amounts for low-performance damages are
 	
    amount   performance level
 	
    . . . .   for   . . . .
 	
Therefore:
 	
▪ An amount for low-performance damages of £ . . . . . . is included in the amount due assessed to be paid by the *Contractor*	

NEC4: The Role of the *Project Manager*
ISBN 978-0-7277-6353-2

ICE Publishing: All rights reserved
http://dx.doi.org/10.1680/nectrpm.63532.171

Index

abbreviations, xiii
Accepted Programme
  in general, 38–39
  access, 50
  amount due, 48
  compensation events, 41, 150
  Contractor, 24, 38–40
  early warning procedure, 50
  early warnings, 40–41
  first, 24
  other procedures, 40
  payment, 41
  Project Manager, 24, 38–41, 59–60, 103
  Scope, 59–60
  wallchart, 60
  see also programmes
access
  Accepted Programme, 50
  programmes, 46
  works, 35, 36, 74, 77–78, 132, 146, 155–156
access date, 100, 130, 131, 132, 145, 146
Activity Schedule, 47
Adjudicator, 23, 64
agendas
  in general, 29
  Defects meetings, 125–126
  early warning meetings, 127–128
  first meeting Supervisor, 109
  monthly meetings, 120, 143
  weekly meetings, 119
agreed procedures, 21–22, 23
amount due
  Accepted Programme, 48
  assessment
    after Completion, 157, 159–160
    checklists, 137–139, 155–156, 159–160, 169
    Client, 62, 155, 159
    Contractor, 47–49, 60, 62, 135–136, 138, 153, 155–156, 159–160
    final, 82, 159–160, 169
    first, 46–49, 135–139
    later months, 60, 62–63, 153
    Project Manager, 41, 47–49, 60, 62–63, 82, 135–136, 153, 155–156, 159
  Bill of Quantities, 48, 135, 137
  compensation events, 48

costs not included in, 24
appointing, *Project Manager*, 8
assessment
  amount due. *see* amount due
  Completion, 71

behaviours, 12
Bill of Quantities, 48, 62, 135, 137, 138, 153
bonus, early completion, 12, 70–71, 74, 161, 162, 163

checklists
  assessment amount due
    after Completion, 159–160
    final, 169
    first, 137–139
    outwith core clause section 5, 155–156
  Client, 99–102
  Completion, 161–163
  contract documents, 103–104
  Defects Certificate, 165
  defects date, 165
  early warnings, 141
  information collection assessment amount due
    after Completion, 157
    final, 167
    first, 135–136
    later months, 153
  kick-off meeting, 111–113
  programmes, 131–133, 145–147
  quality management, 123
  Scope/Contract Data, 105–107
  submitted programmes, 149–151
  take over, 162, 163
  templates *Project Manager*, 115–117
Client
  agreed procedures, 21–22
  amount due, 62, 155, 159
  Bill of Quantities, 48
  communication system, 11
  Completion, 70–71
  contract strategy, 17
  Defects, 70, 77–78
  Defects Certificate, 82
  programme, 18, 45–46
  Project Manager, 7, 10, 12, 23, 93–94, 96, 99–102
  project team setup, 21

Client (*continued*)
quality management, 35–36
subcontracting, 32
take over, 72–74, 78
tests/inspections, 63
Working Areas, 24
collaborative working, 12–13
communications
Contractor, 30
pro forma, 11
Project Manager
in general, 10, 11, 23
overview, 91–98
quality management, 35–36
rules, 91
templates, 29
protocol, 11, 64, 111
system, 11
templates, 29–30
see also notification
compensation events
Accepted Programme, 41, 150
amount due, 48
Contractor, 48, 52
deemed acceptance, 54
early warnings, 52
links to other clauses, 51
notification, 52–54
procedure
decisions, 53–54
deemed acceptance, 54
key stages, 52
Project Manager, 48, 52–54
sources of information, 49
tests/inspections, 63
timelines, 51
Completion
in general, 69
assessment, 71
checklist, 161–163
consequences of, 70–71
Contractor, 12, 70–71, 74, 161, 162, 163
Defects, 69–71, 77–78
links between Completion Date, take over, 72–74
Project Manager, 69–71
Scope, 69–70
sectional, 70, 138, 151, 161–162
tests/inspections, 63
X Options, 70–71
see also Completion Date; take over
Completion Date
in general, 69–71
links between Completion, take over, 72–74
Price Adjustment Factor, 70, 163
Scope, 70
see also Completion; *completion date*; date of
Completion
completion date, 46, 69
see also Completion Date; date of Completion

contract, acting as stated in, 30, 64
Contract Data
Early Warning Register, 38
optional statements, 17
Project Manager, 8, 18, 105–106
contract documents
checklist, 103–104
Project Manager, 18
contract strategy, 17
Contractor
Accepted Programme, 24, 38–40
access to *works*, 132, 146, 155–156
amount due,
final, 159–160
first, 47–49, 135–136, 138
later months, 60, 62, 153
outwith core clause section 5, 155–156
communications, 30
compensation events, 48, 52
Completion, 12, 30, 70–71, 74, 161, 162, 163
Defects, 77–78
Defects Certificate, 82
Defined Cost, 24, 49, 50, 60, 96, 98
design, 23–24, 31–32
early involvement, 71, 98, 160
early warnings
failure to provide, 51
procedure, 50
meetings, 120
payment, 41
programmes, 45–46, 59, 149–151
Project Manager, 7, 10, 12, 25, 38, 47–49, 50–55, 91–98
proposed instruction, 55
quality management, 37, 64, 112
subcontracting, 32–33
take over, 74
Working Areas, 24
cost consultant (quantity surveyor), 22

date of Completion, 69–70
deemed acceptance, 54
defect correction period, 77, 78, 81, 82, 125, 155
Defects
Client, 70, 77–78
Completion, 69–71, 77–78
Contractor, 77–78
meetings, 31, 125–126
notification, 36
Others, 70
Project Manager, 9, 22–23, 70, 77–78
Supervisor, 9, 22–23, 70
see also defect correction period; Defects Certificate;
defects date; quality management; tests/inspections
Defects Certificate
in general, 81–82
checklist, 165
Client, 82
Contractor, 82
Supervisor, 81–82

defects date
in general, 70, 77, 78, 81–82
checklist, 165
Defined Cost
calculating, 12
Contractor, 24, 49, 50, 60, 96, 98
not included in, 24
Project Manager, 49, 53, 60
delay damages, 71, 74, 159
delegation, 21, 22
design
in general, 23–24
Equipment, 32
works, 31–32
Dispute Avoidance Board, 64, 65
dispute resolution
Project Manager, 64–65
W1, W2 and W3 options, 65
documents, 18, 32, 103–104

Early Warning Register
in general, 7
Contract Data, 38
matters included, 36, 38
Project Manager, 36, 38
early warnings
Accepted Programme, 40–41
checklist, 141
compensation events, 52
meetings, 38, 127–128
notification, 36, 38, 49, 50
procedures, 7, 50
Project Manager, 50–51, 141
Equipment
marking, 55
quality management, 64
experience, learning from, 86

first programmes, 18, 45–46, 59, 129–133
see also Accepted Programme; programmes

Guidance Notes, 30, 49, 70, 74

inflation, price adjustment, 70, 138, 167
information
collected, before assessment amount due, 135–136, 153, 157, 167
included in
first programmes, 46
revised programmes, 59
subsequent programmes, 145–147
sources, compensation events, 49

kick-off meeting, 25, 111–113

management meetings, 30, 31
management procedures, 30, 31
see also meetings
marking, 55

meetings
agendas, 29, 109, 119–120, 125–126, 127–128, 143
Defects, 31, 125–126
early warnings, 38, 127–128
kick-off, 25, 111–113
management, 30, 31
monthly, 29, 31, 45, 120, 128, 143
post-project evaluation, 85–86
regular, 31, 45, 119–120
Supervisor, 109, 143
weekly, 31, 45, 119
monthly meetings, 29, 31, 45, 120, 128, 143
mutual trust and cooperation, 30, 64

naming, *Project Manager*, 8
notification
in general, 11
agreed procedures, 23
compensation events, 52–54
Defects, 36
early warnings, 36, 38, 49, 50
templates, 21, 30
see also communications

Others
Defects, 70
meetings, 120

payment
Accepted Programme, 41
procedures, 63
period for reply
in general, 91–94, 96–98, 99
extension, 51
planner (programmer), 22
Plant/Materials
marking, 55
quality management, 64
post-project evaluation, 85–86
Price Adjustment Factor, 70, 163
price adjustment for inflation, 70, 138, 167
Price for Work Done to Date, 48–49, 62
procedures
agreed, 21–22, 23
compensation events
decisions, 53–54
deemed acceptance, 54
key stages, 52
early warnings, 7, 50
management, 30, 31 *see also* meetings
payment, 63
programmer (planner), 22
programmes
accepting, 38–40
access, 46
Client, 18, 45–46
completion date, 46
Contractor, 45–46, 59, 149–151
first, 18, 45–46, 59, 129–133

programmes (*continued*)
  information included, 46, 131–133, 145–147
  look ahead, 59
  Project Manager, 18, 39, 46, 59
  revised, 59
  sources of information, 46
  submitting, 45, 149–151
  see also Accepted Programme
project management triangle, 7
Project Manager
  Adjudicator, 23, 64
  appointing, 8
  behaviours, 12
  Client, 7, 10, 12, 23, 93–94, 96, 99–102
  Contract Data, 105–106
  Contractor, 7, 10, 12, 25, 38, 47–49, 50–55, 91–98
  Contractor's design, 31–32
  good and bad habits, 13
  meeting agendas. *see* agendas
  meetings. *see* meetings
  naming, 8
  other project members, 22–23
  planner/programmer, 22
  quantity surveyor, 22
  role/duties. *see* role/duties
  Scope, 105–107
  Senior Representatives, 64
  subcontracting, 32–33
  Supervisor, 10, 22–23, 33–36, 70
  in *Supervisor*'s role, 30–31
  templates, 115–117
  Working Areas, 24
project team
  in general, 30
  meetings, 120
  setup, 21

quality management
  aspects in project, 64
  checklist, 123
  Contractor, 37, 64, 112
  Project Manager
    in general, 33
    access to *works*, 35
    approaches, 36
    communications, 35–36
    considerations, 121–122
    costs assessment, 34–35
    levels of involvement, 37
  Supervisor, 33–34, 35–36, 121–122
  system, 37, 64, 112
  see also Defects
quantity surveyor (cost consultant), 22

regular meetings, 31, 45, 119–120
  see also monthly meetings; weekly meetings
retention (Option X16), 17, 62, 71, 74, 82, 139, 162, 163, 169
revised programmes, 59
role/duties (*Project Manager*)

general, 7, 8–9, 29
Accepted Programme, 24, 38–41, 59–60, 103
access to *works*, 35, 36, 74, 77–78, 132
Activity Schedule, 47
after contract before start, 21–25
agreed procedures, 21–22
amount due assessment
  in general, 41
  final, 82, 159
  first, 47–49, 135–136
  later months, 60, 62–63, 153
  outwith core clause section 5, 155–156
communications
  in general, 10, 11, 23
  overview, 91–98
  quality management, 35–36
  rules, 91
  templates, 29
compensation events, 48, 52–54
Completion, 69–71
Contract Data, 8, 18
contract documents, 18, 103–104
contract strategy, 17
Contractor's design, 23–24
costs assessment, 34–35
Defects, 9, 22–23, 70, 77–78
Defined Cost, 49, 53, 60
delegation, 21, 22
dispute resolution, 64–65
Early Warning Register, 36, 38
early warnings, 50–51, 141
instructing *Contractor*, 55
kick-off meeting, 25, 111–113
marking, 55
post-project evaluation, 85–86
prior to contract, 17–18
programmes, 18, 39, 46, 59
project team, 21, 30
proposed instruction, 55
quality management
  in general, 33
  access to *works*, 35
  approaches, 36
  communications, 35–36
  considerations, 121–122
  costs assessment, 34–35
  levels of involvement, 37
Scope, 18, 105–107
starting on site, 29–41
take over, 71–74
tender evaluation, 17
v. role *Supervisor*, 9

Scope
  Accepted Programme, 59–60
  communication system, 11
  Completion, 69–70
  Completion Date, 70
  Project Manager, 18, 105–107

sectional Completion, 70, 138, 151, 161–162
Senior Representatives, 64
subcontract documents, 32
subcontracting, 64
Subcontractor
  in general, 32
  proposed post-award, 33
Supervisor
  Defects, 9, 22–23, 70
  Defects Certificate, 81–82
  marking, 55
  meetings, 109, 143
  Project Manager, 10, 22–23, 33–36, 70
  Project Manager as, 30–31
  quality management, 33–34, 35–36, 121–122
  tests/inspections, 9, 63
  v. role *Project Manager*, 9

take over
  in general, 69, 72–74
  certificate, 74
  checklist, 162, 163
  Client, 72–74, 78
  consequences of, 74
  Contractor, 74
  Project Manager, 71–74
  X Options, 74
templates

communication, 29–30
  notification, 21, 30
  Project Manager, 115–117
tender evaluation, 17
tests/inspections
  effects on project, 36, 63
  Supervisor, 9, 63
  see also Defects; quality management

W1, W2 and W3 options, 65, 102
wallchart
  Accepted Programme, 60
  dates, 103
  outline programme, 30
  showing progress, 61
weekly meetings, 31, 45, 119
Working Areas, 24, 111
workmanship, 64
works
  access, 35, 36, 74, 77–78, 132, 146, 155–156
  design, 31–32

X1 Option, 70, 138, 167
X5 Option, 70, 138, 151, 161–162
X6 Option, 12, 70–71, 74, 161, 162, 163
X7 Option, 71, 74, 159
X16 Option, 17, 62, 71, 74, 82, 139, 162, 163, 169
X22 Option, 71, 98, 160